Lecture Notes in Computer Science 12103

More information about this series at http://www.springer.com/series/7407

Juan Romero · Anikó Ekárt ·
Tiago Martins · João Correia (Eds.)

Artificial Intelligence in Music, Sound, Art and Design

9th International Conference, EvoMUSART 2020
Held as Part of EvoStar 2020
Seville, Spain, April 15–17, 2020
Proceedings

 Springer

Editors
Juan Romero (iD)
University of A Coruña
A Coruña, Spain

Tiago Martins (iD)
University of Coimbra
Coimbra, Portugal

Anikó Ekárt (iD)
Aston University
Birmingham, UK

João Correia (iD)
University of Coimbra
Coimbra, Portugal

ISSN 0302-9743   ISSN 1611-3349   (electronic)
Lecture Notes in Computer Science
ISBN 978-3-030-43858-6   ISBN 978-3-030-43859-3   (eBook)
https://doi.org/10.1007/978-3-030-43859-3

LNCS Sublibrary: SL1 – Theoretical Computer Science and General Issues

This Springer imprint is published by the registered company Springer Nature Switzerland AG
The registered company address is: Gewerbestrasse 11, 6330 Cham, Switzerland

Preface

The 9th International Conference on Artificial Intelligence in Music, Sound, Art and Design (EvoMUSART 2020) took place during April 15–17, 2020, in Seville, Spain, as part of EvoStar, the leading European event on bio–inspired computation.

Following the success of previous events and the importance of the field of computational intelligence, specifically, evolutionary and biologically inspired (artificial neural network, swarm, alife) music, sound, art, and design, EvoMUSART has become an EvoStar conference with independent proceedings since 2012.

Although the use of artificial intelligence (AI) for artistic purposes can be traced back to the 1970s, the use of AI for the development of artistic systems is a recent, exciting, and significant area of research. There is a growing interest in the application of these techniques in fields such as: visual art and music generation, analysis, and interpretation; sound synthesis; architecture; video; poetry; design; and other creative tasks.

The main goal of EvoMUSART 2020 was to bring together researchers who are using AI techniques for artistic tasks, providing the opportunity to promote, present, and discuss ongoing work in the area. As always, the atmosphere was fun, friendly, and constructive.

EvoMUSART has grown steadily since its first edition in 2003 in Essex, UK, when it was one of the Applications of Evolutionary Computing workshops. Since 2012 it has been a full conference as part of the EvoStar co-located events.

EvoMUSART 2020 received 31 submissions. The peer-review process was rigorous and double-blind. The international Program Committee, listed below, was composed of 60 members from 19 countries. EvoMUSART continued to provide useful feedback to authors: among the papers sent for full review, there were on average three reviews per paper. The number of accepted papers was 12 long talks (38.7% acceptance rate) and 3 posters accompanied by short talks, meaning an overall acceptance rate of 48.39%.

As always, the EvoMUSART proceedings cover a wide range of topics and application areas, including generative approaches to music and visual art, deep learning, and architecture. This volume of proceedings collects the accepted papers.

As in previous years, the standard of submissions was high, and good quality papers had to be rejected. We thank all authors for submitting their work, including those whose work was not accepted for presentation on this occasion.

The work of reviewing is done voluntarily and generally with little official recognition from the institutions where reviewers are employed. Nevertheless, professional reviewing is essential to a healthy conference. Therefore we particularly thank the members of the Program Committee for their hard work and professionalism in providing constructive and fair reviews.

EvoMUSART 2020 was part of the EvoStar 2020 event, which included three additional conferences: EuroGP 2020, EvoCOP 2020, and EvoApplications 2020. Many people helped to make this event a success.

We thank the invited keynote speakers, José Antonio Lozano (University of the Basque Country, Spain) and Roberto Serra (University of Modena and Reggio Emilia, Italy) for their inspirational talks.

We thank SPECIES, the Society for the Promotion of Evolutionary Computation in Europe and its Surroundings, for its sponsorship.

We thank the local organizing team lead by Francisco Fernández de Vega (University of Extremadura, Spain) and Federico Divina (University Pablo de Olavide, Spain), and also University Pablo de Olavide in Sevilla, Spain, for supporting the local organization.

We thank João Correia (University of Coimbra, Portugal) for the EvoStar publicity, website, and social media service; and also Sérgio Rebelo (University of Coimbra, Portugal) for his important graphic design work.

Finally, and above all, we would like to express our most heartfelt thanks to Anna I Esparcia-Alcázar (Universitat Politècnica de València, Spain), for her dedicated work and coordination of the event. Without her work, and the work of Jennifer Willies in the past years, EvoStar would not enjoy its current level of success as the leading European event on bio–inspired computation.

April 2020

Juan Romero
Anikó Ekárt
Tiago Martins
João Correia

Organization

Conference Chairs

Juan Romero   University of A Coruña, Spain
Anikó Ekárt   Aston University, UK

Publication Chair

Tiago Martins   University of Coimbra, Portugal

Local Chairs

Francisco Fernández   University of Extremadura, Spain
  de Vega
Federico Divina   University Pablo de Olavide, Spain

Publicity Chair

João Correia   University of Coimbra, Portugal

Conference Administration

Anna I Esparcia-Alcazar   EvoStar Coordinator

Program Committee

Mauro Annunziato	ENEA, Italy
Daniel Ashlock	University of Guelph, Canada
Peter Bentley	University College London, UK
Eleonora Bilotta	Università della Calabria, Italy
Tim Blackwell	Goldsmiths College, University of London, UK
Andrew Brown	Griffith University, Australia
Adrian Carballal	University of A Coruña, Spain
Amilcar Cardoso	University of Coimbra, Portugal
Vic Ciesielski	RMIT, Australia
João Correia	University of Coimbra, Portugal
Pedro M. Cruz	Northeastern University, USA
Palle Dahlstedt	Göteborg University, Sweden
Eelco den Heijer	Vrije Universiteit Amsterdam, The Netherlands
Alan Dorin	Monash University, Australia
José Fornari	NICS, Unicamp, Brazil
Marcelo Freitas Caetano	CIRMMT, McGill University, Canada

Philip Galanter	Texas A&M University, USA
Pablo Gervás	Universidad Complutense Madrid, Spain
Gómez-Navarro	
Andrew Gildfind	Google Inc., Australia
Carlos Grilo	Instituto Politécnico de Leiria, Portugal
Andrew Horner	University of Science and Technology, Hong Kong
Takashi Ikegami	University of Tokyo, Japan
Colin Johnson	University of Kent, UK
Daniel Jones	Goldsmiths College, University of London, UK
Anna Jordanous	University of Kent, UK
Amy K. Hoover	University of Central Florida, USA
Maximos	Aristotle University of Thessaloniki, Greece
Kaliakatsos-Papakostas	
Cristobal Kubli	University of Texas at Dallas, USA
Matthew Lewis	Ohio State University, USA
Yang Li	University of Science and Technology Beijing, China
Antonios Liapis	University of Malta, Malta
Alain Lioret	Paris 8 University, France
Roisin Loughran	University College Dublin, Ireland
Penousal Machado	University of Coimbra, Portugal
Roger Malina	International Society for the Arts, Sciences and Technology, USA
Bill Manaris	College of Charleston, USA
Tiago Martins	University of Coimbra, Portugal
Jon McCormack	Monash University, Australia
Eduardo Miranda	University of Plymouth, UK
Nicolas Monmarché	University of Tours, France
Marcos Nadal	Universitat de les Illes Balears, Spain
Michael O'Neill	University College Dublin, Ireland
Philippe Pasquier	Simon Fraser University, Canada
Alejandro Pazos	University of A Coruña, Spain
Somnuk Phon-Amnuaisuk	Brunei Institute of Technology, Malaysia
Douglas Repetto	Columbia University, USA
Nereida	University of A Coruña, Spain
Rodriguez-Fernandez	
Brian Ross	Brock University, Canada
Jonathan E. Rowe	University of Birmingham, UK
Artemis Sanchez Moroni	Universidade Estadual de Campinas (UNICAMP), Brazil
Antonino Santos	University of A Coruña, Spain
Iria Santos	University of A Coruña, Spain
Marco Scirea	IT University of Copenhagen, Denmark
Daniel Castro Silva	University of Coimbra, Portugal
Benjamin Smith	Indianapolis University and Purdue University, USA
Gillian Smith	Northeastern University, USA
Stephen Todd	IBM, UK

Paulo Urbano   Universidade de Lisboa, Portugal
Anna Ursyn   University of Northern Colorado, USA
Dan Ventura   Brigham Young University, USA

Contents

A Deep Learning Neural Network for Classifying Good and Bad Photos

Stephen Lou Banal$^{(\boxtimes)}$ and Vic Ciesielski$^{(\boxtimes)}$

School of Science, RMIT University, GPO Box 2476, Melbourne, VIC 3000, Australia
stephen.banal@gmail.com, vic.ciesielski@rmit.edu.au

Abstract. Current state-of-the-art solutions that automate the assessment of photo aesthetic quality use deep learning neural networks. Most of these networks are either binary classifiers or regression models that predict the aesthetic quality of photos. In this paper, we developed a deep learning neural network that predicts the opinion score rating distribution of a photo's aesthetic quality. Our work focused on finding the best pre-processing method for improving the correlation between ground truth and predicted aesthetic rating distribution of photos in the AVA dataset. We investigated three ways of image resizing and two ways of extracting regions based on salience. We found that the best pre-processing method depended on the photos chosen for the training set.

Keywords: Deep learning · Aesthetics · Photography

1 Introduction

Aesthetic assessment is an inherently complex task to mimic due to its subjective nature. In professional photography, there are several techniques used by photographers to create photos of good aesthetic value, for example, color harmony [18], composition style (e.g. landscape) and subject arrangement [6,13]. Yet, current state-of-the-art deep learning networks that classify photos as good or bad still have limitations in generalizing on various photography challenges. This is mainly due to the sheer complexity of features, opinionated nature of this task and several extraneous factors like context, emotion, and prior knowledge [10]. This can be observed in the aesthetic visual analysis (AVA) benchmark dataset where ratings of photos judged on several photography challenges have high score variance [17]. An example of this is shown Fig. 1a which has a high mean score of 8.57 while Fig. 1b received a low mean score of 3.11 even though both have a good composition style. One could reason that Fig. 1b received a low score since it did not meet the *Live Music* challenge criteria. Binary classification of aesthetic quality is not sufficient to interpret aesthetic quality. It is more robust to know other statistical properties, like mean and standard deviation, to understand if there is consensus or disparity in opinions. Figure 1a and c

© Springer Nature Switzerland AG 2020
J. Romero et al. (Eds.): EvoMUSART 2020, LNCS 12103, pp. 1–16, 2020.
https://doi.org/10.1007/978-3-030-43859-3_1

shows this observation. Lastly, an explanation that justifies a rating is sometimes necessary. For example, Fig. 1c has a mean score of 6.67 even though it is incomprehensible for most people.

(a) Free Study 8.57 (±1.46)
(b) Live Music 3.11 (±2.08)
(c) Motion Blur 6.67 (±2.13)

Fig. 1. Sample photos submitted on specific photography challenges and their ratings.

Recent breakthroughs in deep learning have made significant progress in complex computer vision tasks. In aesthetic assessment, the state-of-the-art deep learning networks have outperformed conventional machine learning methods in terms of predictive accuracy. Several techniques are used to achieve the state-of-the-art, like multi-patch selection [11,14], transfer learning [15], various multi-subnet configurations [2,24], and custom objective functions [22]. Common to all these solutions is a pre-processing step where images are transformed into a format that can be consumed by a neural network. The usual approach is to resize a photo to a smaller dimension which in most scenarios causes image distortion [2,11,22]. The characteristic of most aesthetic models makes it difficult to correlate a prediction to the original photo. Also, the aesthetic qualities of photos are not assessed based on the original photo. Few studies focus on whether the whole image or only specific salient regions of an image is important to evaluate its aesthetic quality and whether such pre-processing steps affect accuracy. Another characteristic of existing work is most are binary classifiers and do not use the richer human opinion score distribution data provided in the AVA dataset. Recent work by [22] called NIMA has shown that a simple architecture that predicts rating distribution can achieve accuracy that is comparable to the state state-of-the-art and close to human perception. In this work, earth mover distance (EMD), defined in Sect. 3.1, is used as a loss function which has been shown to have outperformed regression and cross-entropy based CNNs in aesthetic evaluation tasks [7]. Inspired by this work, this research developed a deep learning network that predicts aesthetic quality rating distribution of photos. In addition, we improve upon the work of the state-of-the-art by studying various image pre-processing methods that best improve the correlation of predicted and ground truth distribution of the ratings.

1.1 Research Questions

1. Which *pre-processing* methods best improve the correlation of predicted and ground truth aesthetic rating distributions?
2. Does focusing on the *salient* regions of a photo improve the correlation of predicted and ground truth aesthetic rating distributions?

2 Related Work

2.1 Aesthetics and Photography

Aesthetics is defined as "a set of *principles* concerned with the *nature* and *appreciation of beauty*" [1]. In photography, a photo with good aesthetic quality follows certain principles in terms of color (e.g. color harmony), composition rules (e.g. rule of thirds) and style (e.g. macro). These qualities can be grouped into low level and high-level features. Low-level features describe the pixel properties of a photo (e.g. color harmony, brightness) [9]. On the other hand, high-level features of a photo describe its composition style (e.g. landscape) and subject (e.g. portrait) [6,13]. Several techniques have been developed to automate the assessment of these aesthetic features using both *computational approaches* and *deep learning* neural networks.

2.2 Aesthetic Neural Network Architecture

Current state-of-the-art aesthetic assessment networks vary on key elements: *network structure*, *base architecture* and *objective function*. In terms of structure these networks can be single-column [22], multi-column [11,21] and multi-staged [23,24]. Most of these networks are single tasked that produce binary or regression output. Others are multi-tasked which predict not only aesthetic quality but also other photo descriptors like category and style [2,5]. Recent state-of-the-art inherits pre-trained object detection architecture. For example InceptionV2 and VGG16. While these networks are trained for a different purpose a technique called *transfer learning* is used by aesthetic classifiers to re-use the learned image features of these networks. For example, [15] fine-tuned Places205-GoogLeNet3 with its scene categorization CNN to classify seven pre-defined scenes. On the other hand, several architectures, like [22], use existing object detection models as its base neural network and appends its own fully-connected layer to output a 10-neuron softmax to predict rating distribution. In the work of [21], VGG16 is used in its multi-patch subnet to extract deep features of photos using multiple patches. Statistical properties of a data set affect the predictive accuracy of machine learning models. In the AVA dataset, there are more than twice as many good photos as bad photos. To circumvent these issues, several aesthetic neural networks created custom objective functions. For example, NIMA [22] models the problem by predicting the rating distribution of a photo. To achieve this, EMD is used as a loss function. UNNA [2] also adopted the same objective function and have concluded it is effective in improving accuracy. BDN [24] on the other hand, used Kullback-Leibler (KL) divergence as its loss function to predict rating distribution. MP-Net [21] uses several objective function MP_{ada}, MP_{avg} and MP_{min} which are designed to giving more weights to misclassified patches of photos.

2.3   Deep Learning and Feature Extraction Approaches

In general, the *color, subject, composition style* and *semantic* features of a photo define its aesthetic quality. Several researchers have observed that simple color themes, certain color pairs, and gradients are perceived as aesthetically appealing [9,18,19]. In the computational approach, these features are extracted using methods like color histogram and luminance distribution [4]. In deep learning, color features are learned through a series of convolutional layers. Images are presented to the network as matrices of RGB values with their corresponding labels. Through back-propagation, the network weights are updated which approximate a function that can assess a photo's low-level qualities. One of the goals of photography is to capture a subject in a given thematic setting. Extracting the subject, therefore, is one of the elements required to assess aesthetic quality. In computational approaches, extracting subject feature requires specialized algorithms. For instance, [13] use a unique subject area extraction per scene category. For example, a face detection algorithm is used for *human* features. Other approaches, use salience maps to extract the important pixels. In deep learning, subject feature extraction relies on the ability of a network to learn features of the subject in an image. Given that photos usually are larger than the image input size requirement of these networks, a set of patch selection techniques were devised. For instance, [12] developed a neural network that uses randomly selected patches to train a multi-patch aggregation network. In the work of [23], a multi-staged deep cropping algorithm is used to select the best regions for aesthetic classification. This network first learns viable candidate patches that are visually important using its *attention box prediction* network and then uses the candidate patches as input to a *aesthetic assessment* network to produce a classification. In *A-Lamp* [14], its network is made up of multi-patch and layout aware subnets used to select the most discriminative and informative patches in the input image. The attention-based objective function is developed by [21] to improve the learning efficiency of the network and also address the class imbalance in the AVA dataset. *Composition* involves techniques used to arrange elements in a photo like the rule of thirds. In computational approaches, features pertaining to layout are extracted using mathematical models that identify the position of salient regions in a scene. In the work of [26], salient regions of a photo are used to model how a scene of a photo is viewed. In this work probabilistic models are used to mimic how humans view salient regions of a photo. In deep learning, the extraction of composition styles relies mostly on using the network to learn high-level features of images like the layout. For example, RAPID [11] created a double-column network where global and local views of an image are used to evaluate aesthetic quality. DMA-Net [12] attempts to extract composition style by using multiple patches of a photo to train a multi-patch aggregation network. *MNA-CNN* [15] have shown that a photo's aesthetic quality is affected by image resize due to resolution reduction and distortion. To address this problem, an adaptive spatial pyramid pooling layer is used for the network to accept images of arbitrary sizes. Other approaches use a custom representation to extract the composition features, like [14], where its layout aware subnet extract layout of detected salient objects in a photo. A&C CNN [8] approached the problem by grouping photos into three categories, namely: *scene*, *object* and *texture*. Each has its own CNN aimed at extracting features of a specific category: for scenes, warping and padding are used to get the global view of a photo; for objects, salience is used to extract the subject; and for texture, patches are extracted based on intensity levels of nearby pixels and similarity of sub-images in a photo.

3   Baseline Architecture

Our deep learning network uses Inception-V2 as its core architecture and EMD loss as the objective function. Inception-V2 is selected as base architecture to compare our results with the best model configuration published in NIMA [17] paper which uses the same network architecture. The last layer of this network is a fully-connected layer followed by a soft-max activation layer with a 10 neuron output. The output represents the 10-scale rating distribution prediction of a photo. Compared to [17] our implementation is trained using different pre-processing algorithms. For details of each layer see Fig. 2.

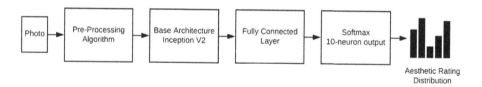

Fig. 2. Baseline deep learning network.

3.1   Objective Function, Optimizer and Hyper Parameters

In our baseline architecture, earth mover distance (EMD) squared loss is used as the objective function. EMD loss is defined as:

$$EMD(p, \hat{p}) = (\frac{1}{N} \sum_{k=1}^{N} \|CDF_p(k) - CDF_{\hat{p}}(k)\|^r)^{\frac{1}{r}} \tag{1}$$

where $r = 2$ and N denotes the number of score buckets. $CDF_p(k)$ is the cumulative distribution function defined as $\sum_{i=1}^{k} p_{s_i}$ where $p = [p_{s_1}, ..., p_{s_N}]$

is the probability distribution of ratings and s_i denotes the score on the ith bucket. Based on predicted probability distribution of rating, the mean μ can be calculated as $\sum_{i=1}^{N} s_i \times p_{s_i}$ while standard deviation σ can be calculated as $\sum_{i=1}^{N}(s_i - \mu)^2 \times p_{s_i}$. EMD with $r = 1$ is used to calculate the closeness of predicted and ground truth rating distributions. Linear Correlation Coefficient (LCC) and Spearman's Rank Correlation Coefficient (SRCC) are calculated on both ground truth and predicted mean and standard deviation of scores. For binary classification, photos with $\mu > 5$ is classified as high quality photo otherwise it is classified as low quality photo.

Adam is used as the optimizer with a learning rate of 1×10^{-6} for training and 1×10^{-7} for tuning. Learning rate decay is also applied at a factor of 0.95 every 10 epochs up to a maximum of 50 epochs. Early stopping is applied after 10 epochs if there are no improvements in EMD loss during training. Minimum early stopping is $\Delta = 0.0001$. The batch size used is 32, momentum is 0.9 and drop out of 0.75 is applied to the last fully-connected layer.

3.2 Dataset and Training Setup

The dataset used by this study is the AVA dataset. This dataset is composed of 255, 530 photos. After pre-processing and image validation, only a total of 255, 503 images are used where a total of 180, 834 images are high quality and 74, 668 images are low quality. High quality photos are those with $\mu \geq 5.0$ while low quality photos are those with $\mu < 5.0$, similar to definition defined in [17]. This makes the dataset imbalanced with more than 70.78% of the photos are high quality. To circumvent this imbalance, a left-right image flipping transformation is applied randomly to low-quality photos to balance the number of both classes. Each photo is rated by 78–549 users on a scale of 1 to 10 based on a specific photography challenge. The average number of user ratings per photo is 210. Most of the photo's mean rating is in the range of $4.0 \leq \mu \leq 6.0$ and has standard deviation of $1.0 \leq \sigma \leq 1.5$ as shown in Fig. 3a. However, some photos have large variance regardless of mean specifically on photography challenges which require non-conventional techniques or have a subjective assessment. This dataset also has varied image sizes where the significant majority are between 600–700 pixels on its largest side as shown in Fig. 3b.

To train and test the models the images are split into 80% train, 10% test and 10% validation sets, similar to the work of [17, p. 2411]. We also defined three subsets of images based on a photo's aesthetic rating. This grouping helps us determine if our deep learning network learns in obvious cases like very high-quality photos with the least training cost and time. The first group is *unambiguously rated* photos which are composed of 57,760 images. This group's mean rating is in the range $\mu \leq 4.0$ or $\mu > 6.0$ which indicates that a significant number of users rated the photos as either very good or bad quality, respectively. The second subset is *ambiguously rated* images with a total of 197,743 images. These are photos whose aesthetic ratings are in the range $4.0 < \mu \leq 6.0$. The mean rating close to the median 5.0 indicates that there is no extreme opinion on a photo's aesthetic quality. Lastly, we train our network on *all photos*.

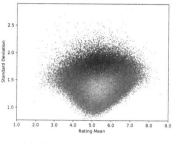

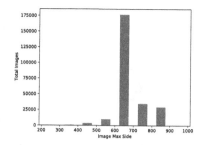

(a) Scatter plot of μ and σ (b) Histogram of Image Maximum Side

Fig. 3. AVA dataset statistics

4 Image Pre-processing Algorithm and Aesthetics Assessment Accuracy

The limitation of commonly used deep learning networks is it has restrictions on the input image size. For example, VGG16 by default only accepts 224 × 224 pixel images. This, therefore, requires users of these networks to implement their pre-processing function that resizes the input images to the correct dimension. Since most state-of-the-art deep learning aesthetic classifiers re-use these networks, this same limitation is inherited by these classifiers. We can define a preprocessing algorithm as a function $\hat{x} = f(x)$ such that when applied to image x its output is a pre-processed image $\hat{x}$. In this study, we want to identify which *image pre-processing algorithm* best improves the correlation of predicted and ground truth rating distributions. We identified and compared three known image pre-processing algorithms: *random*, *global* and *center*.

Random. In this algorithm, each photo is re-sized by shrinking the whole image into a 256 × 256 pixel regardless of the original input image size. The aspect ratio of an image is therefore not preserved if an image is not a square surface. Example photo is shown in Fig. 4b.

Global. In this method, the largest side of the image is resized to 256 pixels and the smaller side is padded with black color. The aspect ratio of a photo is preserved. Example photo is shown in Fig. 4c.

Center. In this method, the smallest side of an image is resized to 256 pixels. The center 256 × 256 patch of the image is then cropped. The aspect ratio of a photo is preserved when this pre-processing method is applied to a photo. Example photo is shown in Fig. 4d.

4.1 Image Pre-processing and Unambiguously Rated Photos

In this experiment, we configured our network using very high and very low-quality photos. These are photos with a mean aesthetic rating in the range of $\mu \leq 4.0$ and $\mu > 6.0$. Our result indicates that *center* produces the best model

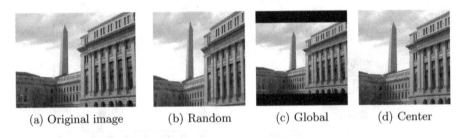

(a) Original image   (b) Random   (c) Global   (d) Center

Fig. 4. Example photo applied with *random, global* and *center* image pre-processing algorithm.

with an accuracy of 93.37%, SRCC of 0.9623 and LCC of 0.4352. Our results also reveal that *random* and *center* have similar linear correlation while SRCC indicates *center* is better than the rest.

The learning rate graphs of each image pre-processing function is shown in Fig. 5. It can be observed that overfitting has occurred at the 20th epoch in both *random* and *global*, while for *center* overfitting occurred at the 25th epoch. Our test show that *random, global* and *center* produced an accuracy of 93.37%, 92.40% and 93.54% respectively. Results also indicate that *center* pre-processing method is good at predicting both low and good quality photos with an F1 score of 0.9632 compared to *random* and *global* which has an F1 score of 0.9623 and 0.9568 respectively with no statistically significant difference. Completing this experiment required us to try out the different number of tunable layers. Initially, we trained using only 10 layers and have achieved at least 80% accuracy on all models. But, the results were disappointing since the model has zero false discovery rate, therefore it is unable to predict low-quality photos. Updating the model to tune all layers during training resulted in a model that can predict both low-quality and high-quality photos.

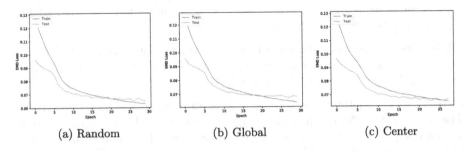

(a) Random   (b) Global   (c) Center

Fig. 5. EMD loss graphs of each of each model using *random, global* and *center* Image Pre-processing algorithm on photos with very high and very low rating.

4.2   Image Pre-processing and Ambiguously Rated Photos

In this experiment, we trained our model using all photos with aesthetic rating mean in the range $4.0 \leq \mu < 6.0$. A total of 197,323 photos are used in this experiment, where 157,859 are for training, 19,733 for validation and 19,731 for tests. Our result shows that models that use *random* as image pre-processing step produce the best result in terms of accuracy, SRCC, and LCC but with no statistically significant difference compared to *global* and *center*. The model that used random produced an accuracy of 68.07% while *global* and *center* produced an accuracy of 67.24% and 67.97% respectively.

4.3   NIMA Model Comparison

To compare our implementation against published NIMA results we trained our network using all photos. We used $204,403$ photos for training, $25,551$ for validation and $25,548$ for tests. All layers are tuned and the learning rate is set to 1×10^{-6}. We intentionally increased the learning rate from 3×10^{-7} to 1×10^{-6} to reduce time and cost to train the network.

Our result show that *center* is the best image pre-processing algorithm followed by *random*. Interestingly, last among the three is *global* with significant difference in accuracy compared to *center* and *random*. Compared to NIMA's 81.51% accuracy our best model, *center*, only reached 77.65% accuracy. SRCC and LCC result of these two models show that both *center* and *random* have almost similar result and is close to NIMA's result as shown in Fig. 6. Also, our test shows that while our best model *center* has an accuracy below NIMA's 81.51% accuracy, it has an F1 score of 85.22%. This gives us insight that our model can predict both high and low-quality photos well with imbalanced classes like the AVA benchmark dataset.

Model Name	Accuracy	F1 Score	EMD	SRCC *Mean*	LCC *Mean*	SRCC *StdDev*	LCC *StdDev*
NIMA (InceptionV2)	81.51%	–	0.0500	0.6120	0.6360	0.2180	0.2330
Baseline + Random	77.19%	0.8485	0.0717	0.6054	0.6149	0.2599	0.2698
Baseline + Global	71.22%	0.8434	0.0728	0.3709	0.3820	0.0832	0.0837
Baseline + Center	77.65%	0.8522	0.0713	0.6022	0.6045	0.2565	0.2566

Fig. 6. Training result of each model trained using *random*, *center* and *global* image pre-processing method using all dataset

Our result reveal that both *random* and *center* are significantly better than *global* as image pre-processing function. We also observed that networks that use *center* as image pre-processing function produces models with better accuracy when trained with photos that have a very high and very low mean aesthetic rating. On the other hand, *random* works well on photos that are ambiguously rated. While our result shows that we were unable to beat NIMA's accuracy, our trained models can achieve almost similar SRCC and EMD as NIMA. In addition, F1 score also indicate that our model performs well on predicting both high and low quality photos with *center* leading at 85.22%, random second at 84.85% and last *global* at 84.34%.

5   Salience and Aesthetic Assessment Accuracy

In this study, we investigated the correlation of a photo's salient regions to its aesthetic quality rating. Our intuition is that users evaluate aesthetic quality using only the salient areas of an image. Therefore, for large photos, the locations which have more salient regions should be used for aesthetic assessment. We developed two algorithms that extract the most salient regions in a photo using *salience merge* or *salience clustering*. In both algorithms, the salient regions in an image are extracted using fine-grained saliency approach [16] shown as $find\_salient\_regions(image)$ method in Algorithms 1 and 2. The extracted image regions are then used to train our deep learning aesthetic assessment networks.

Salience Merge. In this algorithm, salient regions of a photo are identified and then sorted in descending order by area. The top N regions are merged and then selected as the image patch used for aesthetic assessment. Algorithm 1 defines how the salient regions are selected. Figure 7 is an example image showing the original photo, the salient regions identified, the merged region and the final extracted image. In this research, we configured our algorithm to only detect the top $N = 10$ salient regions with a minimum area of 100 pixels. The rationale is to avoid regions that are too small for merging that would cause the algorithm to output an image that is too small for aesthetic assessment. For images where the number of salient regions detected is below the threshold N, the whole image is used instead.

Algorithm 1. Salience Merge algorithm

---

   procedure FIND_REGION(image, max_regions, target_dimension)
       $regions \leftarrow find\_salient\_regions(image)$
       $largest\_regions \leftarrow sort\_regions\_by\_area(regions, max\_regions)$
       $left \leftarrow \min_{left}(largest\_regions)$
       $right \leftarrow \max_{right}(largest\_regions)$
       $bottom \leftarrow \min_{bottom}(largest\_regions)$
       $top \leftarrow \max_{top}(largest\_regions)$
       $region \leftarrow Rect(top, right, bottom, left)$
       return $center\_crop(image, region, target\_dimension)$

Salience Clustering. In this algorithm, the patch used for aesthetic assessment is determined by first identifying the salient regions of a photo and then applying the K-Means clustering algorithm to determine where the salient regions tend

 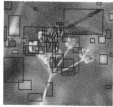  

(a) Original image   (b) Salient regions   (c) Merged regions   (d) Final image

Fig. 7. Example photo applied with salience merge algorithm

Algorithm 2. Salience Clustering algorithm

---

    procedure FIND_REGION(image, max_cluster, top_clusters, target_dimension)
        $regions \leftarrow find\_salient\_regions(image)$
        $clusters \leftarrow k\_means\_cluster(regions, max\_cluster)$
        $largest\_clusters \leftarrow find\_largest\_clusters(clusters, top\_clusters)$
        $left \leftarrow \min_{left}(largest\_clusters)$
        $right \leftarrow \max_{right}(largest\_clusters)$
        $bottom \leftarrow \min_{bottom}(largest\_clusters)$
        $top \leftarrow \max_{top}(largest\_clusters)$
        $region \leftarrow Rect(top, right, bottom, left)$
        return $center\_crop(image, region, target\_dimension)$

---

to concentrate in a photo. The maximum number of clusters is set to $K = 10$. The salient regions of the top N clusters with the largest number of member salient regions are selected and then merged. In our experiments, we set $N = 5$. The merged region defines the patch in the image used for aesthetic assessment. See Algorithm 2 for the details of the algorithm. Figure 8 is an example photo applied with this algorithm.

5.1 Salience Algorithms and Unambiguously Rated Photos

In this experiment, we trained our network using photos with a very high and very low aesthetic rating. We call these dataset as *unambiguously rated* photos where each photo have a mean rating $\mu \leq 4.0$ and $\mu > 6.0$. On this dataset, the model that used *salience merge* produced an accuracy of 91.86% while the model that used *salience clustering* produced an accuracy of 91.82%. Interestingly, the EMD loss of *salience merge* and *salience clustering* did not show any overfitting has occurred as shown in Fig. 9. Our result also shows that while *salience merge* is better than *salience clustering* in terms of accuracy, there is no statistically significant difference in F1 score, SRCC, and LCC.

(a) Original image   (b) Original image   (c) Original image

(d) Original image   (e) Original image   (f) Original image

Fig. 8. Example photo applied with salience clustering algorithm

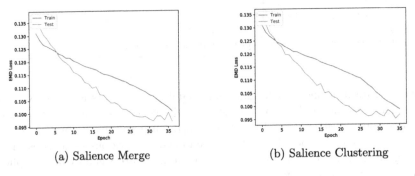

(a) Salience Merge   (b) Salience Clustering

Fig. 9. EMD loss graphs of models trained using *salience algorithms* and photos with very high and very low rating.

5.2   Salience Algorithms and Ambiguously Rated Photos

In this experiment, we trained our network using *ambiguously rated* photos which has mean aesthetic rating in the range $4.0 \leq \mu < 6.0$. There are a total of 197,323 photos used for training the model where 104,509 is used for training, 13,064 for validation and 13,063 for testing. Our result show that there is no significant difference in the accuracy of *salience merge* and *salience clustering* which has an accuracy of 71.55% and 71.41% respectively. The F1 score also show no significant difference between the two which is 0.8011 for *salience merge* while 0.8025 for *salience clustering*. Compared to image pre-processing methods *random*, *global* and *center* our result indicate that both salience based algorithm performs significantly better on *ambiguously rated* photo. The best model in image pre-processing function is *center* and *random* which has an accuracy of 67.97% and 68.07% respectively.

5.3   NIMA Model Comparison

In this experiment, we trained our model using the entire dataset and compared them against NIMA. *Learning rate* is set to 1×10^{-6} at a maximum of 30 epochs and all layers of the network is enabled for tuning.

Model Name	Accuracy	F1 Score	EMD	SRCC Mean	LCC Mean	SRCC StdDev	LCC StdDev
NIMA (InceptionV2)	81.51%	–	0.0500	0.6120	0.6360	0.0434	0.0233
Baseline + SM	77.18%	0.8496	0.0728	0.5846	0.5880	0.2513	0.2547
Baseline + SC	76.90%	0.8482	0.0687	0.5711	0.5724	0.2453	0.2344

Fig. 10. Training result of *salience merge* (SM) and *salience clustering* (SC) using all images of the AVA dataset.

Our results reveal that *salience merge* which has an accuracy of 77.18% outperform models that use *salience clustering* which has an accuracy of 76.90%. We compared these results against image pre-processing methods and concluded that models that use salience based algorithms are significantly better than using *global* as the pre-processing function which has an accuracy of 71.22%. In addition, *salience merge* is comparable to *random* given the very close accuracy and F1 score result of both models. As shown in Fig. 6, models that use random has an accuracy of 77.19% and F1 score of 84.85%. Compared to NIMA's 81.51% accuracy, our best salient based model is *salience merge*. Though it did not outperform NIMA in terms of accuracy, our model shows an 84.96% F1 score which indicates it can predict aesthetic rating on both high and low-quality photos well (Fig. 10).

In this study, we have shown two alternative image pre-processing algorithms, namely *salience merge* and *salience clustering*. We proposed these methods of image pre-processing given the intuition that a human eye evaluates aesthetics only on salient regions of a photo. In all experiments, *salience merge* shows better accuracy and F1 score compared to *salience clustering* but with no statistically significant difference. Our result also show that on images with mean aesthetic rating in the range of $4.0 \leq \mu < 6.0$, *salience clustering* or *salience merge* achieved better accuracy with statistically significant difference when compared to *random*, *center* and *global*. While these algorithms are slightly slower in terms of runtime cost due to the use of K-means algorithm and region merging, there is a potential for improving these algorithms and be used to efficiently evaluate the aesthetic quality of large photos.

6   Conclusion

In this research, we studied five image pre-processing methods, namely *random*, *global*, *center*, *salience merge* and *salience clustering*. To identify the best image pre-processing algorithm, we trained our network using photos of varied aesthetic rating ranges: *unambiguously rated* photos which either have a very high or very low aesthetic rating ($4.0 \leq \mu$ or $\mu > 6.0$), *ambiguously rated* photos which has an aesthetic rating in the range $4.0 < \mu \leq 6.0$, and all photos regardless of rating. We trained our neural network using all these dataset groupings and measured their accuracy and F1 score. Our result reveal that models that use *center*, *random* and *salience merge* as image pre-processing algorithm works best when trained using *all photos* regardless of rating. Models that use *salience clustering* and *salience merge* as image pre-processing algorithm works best when trained using *ambiguously rated* photos while *center*, *random* and *global* works best when trained using *unambiguously rated* photos. Figure 11 is a summary of the accuracy and F1 score of each model using the aforementioned image pre-processing algorithms and dataset groups.

Model Name	Unambiguous		Ambiguous		All Photos	
	Accuracy	F1 score	Accuracy	F1 score	Accuracy	F1 score
Baseline+Center	**93.54%**	**0.9632**	67.97%	0.7964	**77.65%**	**0.8522**
Baseline+Random	93.37%	0.9623	68.07%	0.8000	77.19%	0.8485
Baseline+Global	92.40%	0.9568	67.24%	0.7928	71.22%	0.8434
Baseline+SM	91.86%	0.9536	**71.55%**	**0.8011**	77.18%	0.8011
Baseline+SC	91.82%	0.9535	71.41%	0.8025	76.90%	0.8482

Fig. 11. Accuracy and F1 score of each deep learning network trained using ambiguously rated photos, unambiguously rated photos, and all photos regardless of rating.

6.1   Further Work

The image pre-processing and salience algorithms described in this study are simple yet effective ways to extract aesthetically important regions of a photo. The disadvantage of these algorithms is that the parameters are statically defined and are not learned from a labeled data set. Several studies have successfully used salience detection and deep learning models for object recognition like the work of [25]. Further research of its application on aesthetic assessment could be a potential research topic. Another interesting field in deep learning is *AI explainability* and has been a focus of extensive research recently. In this work, *attribution algorithms* were developed to help users and experts alike to understand and explain the decisions of these complex networks. Examples of these state-of-the-art algorithms are LIME (Local Interpretable Model-Agnostic Explanations) [20] and LRP (Layer-wise Relevance Propagation) [3]. Similar algorithms can also be developed for aesthetic assessment models to help identify regions of a photo that are aesthetically important.

References

1. Aesthetics (2019). https://en.oxforddictionaries.com/definition/aesthetics
2. Abdenebaoui, L., Meyer, B., Bruns, A., Boll, S.: UNNA: a unified neural network for aesthetic assessment. In: 2018 International Conference on Content-Based Multimedia Indexing (CBMI), pp. 1–6. IEEE (2018)
3. Bach, S., Binder, A., Montavon, G., Klauschen, F., Müller, K.R., Samek, W.: On pixel-wise explanations for non-linear classifier decisions by layer-wise relevance propagation. PLoS One **10**(7), e0130140 (2015)
4. Bo, Y., Yu, J., Zhang, K.: Computational aesthetics and applications. Vis. Comput. Ind. Biomed. Art **1**(1), 6 (2018)
5. Chang, K.Y., Lu, K.H., Chen, C.S.: Aesthetic critiques generation for photos. In: Proceedings of the IEEE International Conference on Computer Vision, pp. 3514–3523 (2017)
6. Dhar, S., Ordonez, V., Berg, T.L.: High level describable attributes for predicting aesthetics and interestingness. In: CVPR 2011, pp. 1657–1664. IEEE (2011)
7. Hou, L., Yu, C.P., Samaras, D.: Squared earth mover's distance-based loss for training deep neural networks. arXiv preprint arXiv:1611.05916 (2016)
8. Kao, Y., Huang, K., Maybank, S.: Hierarchical aesthetic quality assessment using deep convolutional neural networks. Sig. Process. Image Commun. **47**, 500–510 (2016)
9. Ke, Y., Tang, X., Jing, F.: The design of high-level features for photo quality assessment. In: 2006 IEEE Computer Society Conference on Computer Vision and Pattern Recognition (CVPR 2006), vol. 1, pp. 419–426. IEEE (2006)
10. Leder, H., Belke, B., Oeberst, A., Augustin, D.: A model of aesthetic appreciation and aesthetic judgments. Br. J. Psychol. **95**(4), 489–508 (2004)
11. Lu, X., Lin, Z., Jin, H., Yang, J., Wang, J.Z.: Rapid: rating pictorial aesthetics using deep learning. In: Proceedings of the 22nd ACM International Conference on Multimedia, pp. 457–466. ACM (2014)
12. Lu, X., Lin, Z., Shen, X., Mech, R., Wang, J.Z.: Deep multi-patch aggregation network for image style, aesthetics, and quality estimation. In: Proceedings of the IEEE International Conference on Computer Vision, pp. 990–998 (2015)
13. Luo, W., Wang, X., Tang, X.: Content-based photo quality assessment. In: 2011 International Conference on Computer Vision, pp. 2206–2213. IEEE (2011)
14. Ma, S., Liu, J., Wen Chen, C.: A-lamp: adaptive layout-aware multi-patch deep convolutional neural network for photo aesthetic assessment. In: Proceedings of the IEEE Conference on Computer Vision and Pattern Recognition, pp. 4535–4544 (2017)
15. Mai, L., Jin, H., Liu, F.: Composition-preserving deep photo aesthetics assessment. In: Proceedings of the IEEE Conference on Computer Vision and Pattern Recognition, pp. 497–506 (2016)
16. Montabone, S., Soto, A.: Human detection using a mobile platform and novel features derived from a visual saliency mechanism. Image Vis. Comput. **28**(3), 391–402 (2010)
17. Murray, N., Marchesotti, L., Perronnin, F.: AVA: a large-scale database for aesthetic visual analysis. In: 2012 IEEE Conference on Computer Vision and Pattern Recognition, pp. 2408–2415. IEEE (2012)
18. Nishiyama, M., Okabe, T., Sato, I., Sato, Y.: Aesthetic quality classification of photographs based on color harmony. In: CVPR 2011, pp. 33–40. IEEE (2011)

19. O'Donovan, P., Agarwala, A., Hertzmann, A.: Color compatibility from large datasets. ACM Trans. Graph. (TOG) **30**(4), 63 (2011)
20. Ribeiro, M.T., Singh, S., Guestrin, C.: Why should i trust you?: Explaining the predictions of any classifier. In: Proceedings of the 22nd ACM SIGKDD International Conference on Knowledge Discovery and Data Mining, pp. 1135–1144. ACM (2016)
21. Sheng, K., Dong, W., Ma, C., Mei, X., Huang, F., Hu, B.G.: Attention-based multi-patch aggregation for image aesthetic assessment. In: 2018 ACM Multimedia Conference on Multimedia Conference, pp. 879–886. ACM (2018)
22. Talebi, H., Milanfar, P.: NIMA: neural image assessment. IEEE Trans. Image Process. **27**(8), 3998–4011 (2018)
23. Wang, W., Shen, J.: Deep cropping via attention box prediction and aesthetics assessment, vol. 2017, pp. 2205–2213. Institute of Electrical and Electronics Engineers Inc. (2017)
24. Wang, Z., Chang, S., Dolcos, F., Beck, D., Liu, D., Huang, T.S.: Brain-inspired deep networks for image aesthetics assessment. arXiv preprint arXiv:1601.04155 (2016)
25. Xiao, F., Deng, W., Peng, L., Cao, C., Hu, K., Gao, X.: Multi-scale deep neural network for salient object detection (2018). https://doi.org/10.1049/iet-ipr.2018.5631
26. Zhang, L., Gao, Y., Zhang, C., Zhang, H., Tian, Q., Zimmermann, R.: Perception-guided multimodal feature fusion for photo aesthetics assessment. In: Proceedings of the 22nd ACM International Conference on Multimedia, pp. 237–246. ACM (2014)

Adapting and Enhancing Evolutionary Art for Casual Creation

Simon Colton[1,2(✉)], Jon McCormack[1], Sebastian Berns[2], Elena Petrovskaya[2], and Michael Cook[2]

[1] SensiLab, Faculty of IT, Monash University, Melbourne, Australia
[2] Game AI Group, EECS, Queen Mary University of London, London, UK
s.colton@qmul.ac.uk

Abstract. Casual creators are creativity support tools designed for non-experts to have fun with while they create, rather than for serious creative production. We discuss here how we adapted and enhanced an evolutionary art approach for casual creation. Employing a fun-first methodology for the app design, we improved image production speed and the quality of randomly generated images. We further employed machine vision techniques for image categorisation and clustering, and designed a user interface for fast, fun image generation, adhering to numerous principles arising from the study of casual creators. We describe the implementation and experimentation performed during the first stage of development, and evaluate the app in terms of efficiency, image quality, feedback quality and the potential for users to have fun. We conclude with a description of how the app, which is destined for public release, will also be used as a research platform and as part of an art installation.

1 Introduction

Researchers/practitioners in generative visual art have made many user-guided evolutionary creativity support tools [22], with focus on the quality, fidelity and variety of the images, generation speed and usage of the tools within art practice. It is often overlooked that the tools themselves are enjoyable to use, with users navigation a space of images through a fluid and responsive user interface; being surprised by unexpected results; collecting and curating sets of images; and repeatedly viewing pleasing images. We address here how to design generative visual art tools primarily for non-experts to have fun with during the creative process, with less concern for the value or utility of the generated images.

Our aim is to produce a creativity support tool for both public release and usage as a research platform, aimed primarily at improving users' well-being through playful creation. In [6] and [7], Compton et al. have driven forward the study of such systems, which they call *casual creators* and describe as:

> ... interactive system[s] which encourage fast, confident, and pleasurable exploration of a possibility space, resulting in the creation or discovery of surprising new artefacts that bring feelings of pride, ownership, and creativity to the users that make them. [7]

© Springer Nature Switzerland AG 2020
J. Romero et al. (Eds.): EvoMUSART 2020, LNCS 12103, pp. 17–34, 2020.
https://doi.org/10.1007/978-3-030-43859-3_2

Compton et al. further specify important points about intended users for casual creators and expected usage patterns. In particular, the system should expect no previous domain knowledge or technical experience, all of the learning must occur in the first few minutes, and the software should provide a good experience, even if the user never spends time to gain mastery.

Casual creation is particularly popular on handheld devices such as smartphones and tablets, as evidenced by hundreds of such apps published on iOS and Android stores. Casual music generation tools are particularly numerous, but focusing on visual art generators, apps include mandala makers, paint by numbers and casual drawing/painting apps, fractal image exploration apps, artistic image filtering and evolutionary art apps. These are supplemented by many ad hoc systems for *one-touch creation*, where repeating a single gesture such as tapping or swiping guides the creative process. Most handheld casual creators use some kind of generative process, ranging from ones which are predictable visually, e.g., simulating paint strokes; through ones which introduce modest surprise, e.g., by non-photorealistic rendering of a user-given image; to ones which introduce entirely new images with high novelty and potential for surprise.

We describe here the development of an evolutionary art casual creator which generates imagery as art material for users to create with. The generative process is particle-based, employing five mathematical functions to *initialise* the particles by placing and colouring them in 2D and RGB spaces, and five mathematical functions to determine how the (x, y, r, g, b) values of the particles are *updated* over a certain number of time steps [16]. Lines and shapes are rendered at the particle positions at each timestep and blurring is regularly applied to introduce depth. We have investigated fitness functions for searching the space and enabling user-guided exploration, and have found various artistic uses for the images produced, e.g., as scene descriptors [4]. Ultimately, however, we realised that the main benefits were in enjoying guiding the generative process, rather than in the value of the images themselves. Hence we felt that a casual creator is an appropriate context for this kind of generative art.

The new app is called *Art Done Quick* and is being initially developed for iOS devices. We have currently completed stage one of the development, producing a minimum viable product that can be put forward for user testing. There has been much adaptation and enhancement of the original generative process, along with user interface design, implementation and experimentation to investigate different ways to embed it in a casual creator. As described in the next section, this has been undertaken within a design methodology that puts user enjoyment first and draws on many design patterns for casual creators identified in [7]. Particular emphasis has been put on instant generation of varied, high quality images as starting points, as described in Sect. 3. We have also drawn heavily on recent advances in machine vision for classifying and analysing generated images, to potentially increase user enjoyment, as described in Sect. 4. Stage two will involve user testing of various hypotheses pertaining to the enjoyment, agency and flow levels that people have when playing with the app. In Sect. 5, we describe some related work, and conclude by highlighting some general points and proposing hypotheses about user enjoyment with Art Done Quick. We also describe how the app, which we aim to be publicly released as an entertainment tool, will also be used in an art installation and as a research platform.

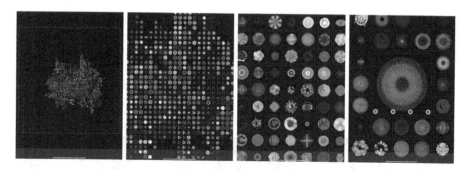

Fig. 1. Screenshots of an Art Done Quick image sheet (iPad Pro version in portrait mode), with 1,000 images showing, depicted at four levels of magnification.

2   A Fun-First Design Methodology

Art Done Quick embeds a human-guided evolutionary art process, with images randomly generated by the above particle-based process, that users can choose to mutate or combine through genomic crossover, producing eight offspring in both cases. Images are generated at 500×500 pixels very quickly, and a user can request that an image be re-rendered at 2000×2000 pixels, which is also fairly fast (see the next section for generation speeds). Users can edit images with post-processing that amount to sequences of image filtering and collaging operations. The post-processing is inherited by an image's offspring, currently without alteration. Our overriding design principle has been to constantly challenge the addition of new functionality or user experience by asking: will people have fun with this? Within this paradigm, there have been three major sources of progress: (a) implementing design patterns arising from the study of casual creators in [7] (b) identifying and minimising frustrating aspects that arise in particular for human-guided evolutionary visual art (c) drawing on the success of existing casual creators for visual art, by implementing similar functionality.

The first design pattern from [7] we adhered to is *no blank canvas*, i.e., minimising the terror of not knowing where to start by giving users artefacts immediately. Related to this is a particular frustration with evolutionary art: it often takes a long time for interesting images to occur. To address this, we designed the app to provide good quality images straight away and have empty boxes inviting taps to instantly fill them with a varied selection of images, as described in the next two sections. Another potential frustration with generative art can be losing valued images, as they have been filed away or otherwise hidden.

To address this, Art Done Quick arranges the generated images to be contained on one large sheet, which can be zoomed in and out of, as per Fig. 1. Tagging and liking images, dropping pins and clustering images in locales enables the user to find images efficiently, and zooming out can aid the memory in finding a particular image or collection. Moreover, having all images present on the sheet lowers the feeling of potentially having lost one. To theme the app, the image sheet is presented as if floating in space, with generated artworks largely circular on a transparent background. The screenshots in Fig. 1 show that this gives an impression of each piece as a floating entity, rather than a rectangular image. The 500 × 500 pixel images are JPEG-compressed in memory, and up to 1,000 can exist on the sheet simultaneously, as shown in Fig. 1.

Fig. 2. Image editing functionality (iPhone version in portrait mode): original image, liquifying options, adding texture, collaging options and adding lighting effects. Below are 8 variations obtained by mutating the final edited image (in the fifth screenshot).

The design pattern in [7] called *mutant shopping* maps directly onto mutation and crossover in Art Done Quick. While fun in general, in evolutionary art, it can often be frustrating if offspring are too similar to the original, as it may feel like little progress has been made. Conversely, if the offspring are too dissimilar to the original, the choice the user made is not reflected in the resulting images, which is usually disappointing. To address this, we derived suitably satisfying balances for mutation and crossover rates, which are fairly conservative at genome level [3]. In work on a different casual creator [12], it was clear that users were frustrated when a generated artefact was nearly perfect, if they couldn't fix the imperfection. For this reason, we added an editing tool to Art Done Quick, so people may change aspects of a generated image. One issue with allowing such editing is that the required improvement might be one of many things, and it is easy to over-compensate by allowing too many editing possibilities. This could complicate the interface and/or introduce choice paralysis – where users are paralysed by an overwhelming number of options to explore [18] – especially if the app begins to feel more like a professional image editing tool. Adhering to the design pattern of *modifying the meaningful*, we made a reasonable number of high-level editing possibilities available, but gave few fine-grained parameters per editing choice. Moreover, as shown in Fig. 2, we kept the editing interface largely homogeneous, to limit feelings of discombobulation.

In designing the editing functionality, we adhered to the *instant feedback* design pattern of [7], so that as the user taps a choice, or changes a value by dragging a slider, the image changes instantly. In the one case where this is not possible (namely changing the original generative parameters and re-rendering at 2000 × 2000 pixels), we adhered to the design pattern of *simulation and approximating feedback*, to give users some indication in advance of what the re-rendered image would look like. Importantly, Art Done Quick offers two tiers of interaction: users can simply use the mutation and crossover mechanisms to explore a space of images without ever feeling the need to alter an image, in which case they would not be exposed to the image editing functionality. This adheres to the design pattern of *limiting actions to encourage exploration*. Mutation is initiated by users double tapping the target original, and they can also drag one image onto a target one, to initiate crossing the two genomes over and producing eight offspring. In both cases, the images around the target move away, leaving eight empty cells to be filled by the generative process.

We experimented with enabling offspring to inherit the edits supplied by the user for an original, and found that it was enjoyable to see these carried over. This enjoyment comes from the realisation that editing an image is actually akin to programming a graphics macro. Moreover, often one of the mutants is more appealing visually than the original, or sometimes strikingly different, while still obviously related – see Fig. 2 for examples. To increase the enjoyment, we decided to extend the editing of images beyond the requirement of merely improving generated pieces, enabling users to take on much more design responsibility, if desired. In choosing what to allow users to edit, we looked at other popular casual creators to see what people seemingly like doing with images: we found that producing mandalas and kaleidoscopes was popular, as was artistic image filtering and the simulation of drawing/painting. To cater for these fun image-based activities, we enabled users to personalise images by adding text and/or importing their own digital photos to overlay as textures or incorporate as stickers in montages. We also implemented a drawing/painting interface (which affords mandalas), and added a number of image filters, each chosen to be fun rather than utilitarian. These include: producing kaleidoscopes and other patterned arrangements of an image; motion blurring; pixelation and making mosaics; liquifying images; adding lighting effects, mapping to colour palettes, adding textures and employing glassy, glowing and other text effects.

The final design pattern we adhered to is entitled *entertaining evaluations*. This advocates for casual creators which evaluate user creations in such a way "that the evaluations can *themselves* be pleasurable and entertaining" [7]. To this end, we have added on-device machine vision techniques to analyse and categorise generated images in order to provide fun feedback, to prompt users to see their creations in a new light. In particular, by pointing out that the machine vision system estimates with high probability that an abstract artwork contains a particular object or portrays a particular scene, Art Done Quick can prompt the user to look at it anew. In doing so, there is often a moment of realisation when the user sees what the app has seen, which can be fascinating. As described in Sect. 4, this is reliable enough to use in making suggestions for image titles and giving users content-based starting points. In Sect. 6, we mention other design patterns to consider in future versions of the app.

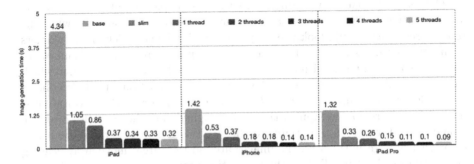

Fig. 3. Image generation speeds for different generative setups for iPad, iPhone and iPad Pro devices. The 1-thread and multi-thread setups used the theming of genomes approach, and produced multiple images simultaneously for average speed increase. Measures were taken to minimise thermal throttling of the devices.

3   Image Generation Efficiency

Random image generation using the original Java version of the evolutionary art system takes around 5 s (on a 2.6 GHz laptop) to produce one 500 × 500 image, of which only around 1 in 50 has any visual appeal/interest. Such slow and pointless generation would be disastrous in a casual creator setting, where instant gratification is an aim. The slow speed of the Java version is due to the trees representing the mathematical functions controlling the particle positions/colours being interpreted at run-time. In a series of user-guided evolutionary sessions [3], we harvested 1,000 genomes (comprising 5 initialisation functions and 5 update functions) which produce varied, interesting and aesthetically pleasing images (subjectively). Each function tree in each of the chosen genomes was flattened into code compilable by Swift (an iOS programming language) and added to a Swift class to be pre-compiled for Art Done Quick. In this way, we extracted 200 initialisation and 200 update functions. Random generation in Art Done

Quick comprises choosing 5 initialisation and 5 update functions from these pre-compiled sets, which gives $200^5 \times 200^5 \approx 10^{23}$ possible combinations. We have found that this gives much visual variation with much reliability, i.e., almost all images produced have some visual interest and aesthetic appeal (subjectively). We found that, with the pre-compilation of the particle functions, producing images on-device was already faster, with 500×500 images produced in less than 1.5 s (on an iPad Pro). The efficiency gains come from the pre-compilation and GPU optimisation of graphics operations on iOS devices.

To further increase generation speed, we experimented with the number of particles and timesteps in the art generation process. More particles for more timesteps means longer rendering times, but does not necessarily produce more interesting images. Originally, we started with a range of 500–1000 particles over 50–100 timesteps for random genomes. We found that changing these ranges to 100–600 and 5–70 respectively had little effect on the visual quality of the images produced, but the *slimmer* generative process was more than four times faster on average, as per Fig. 3. We were also able to simultaneously address speed and quality, while giving users more agency over the random generation. To do this, we derived 20 *themes* which constrain the number of particles, timesteps and shapes used in the generation. Users can choose particular themes, and by default, the app chooses one of the themes randomly, then generates a genome within the theme's constraints. As many themes have below-average particle and timestep coefficients, this improves efficiency a little, as per Fig. 3.

Finally, we took advantage of the multi-core nature of the devices, with generation of five 500×500 pixel images done simultaneously. Generation speeds for this are given in Fig. 3. The devices we tested on were (i) a 6th generation iPad (ii) an iPhone XS, and (iii) a 3rd generation 12.9 in. iPad Pro. The iPhone and iPad Pro would still be considered top of the range, but the iPad would not. We see in Fig. 3 that even on the slowest device (the iPad), the multi-threaded generation using themes with slimmed parameters can produce more than three images per second (with the iPad Pro generating more than 10 per second). To best utilise these rapid generation times, we implemented a caching mechanism, so that Art Done Quick performs random image generation in the background, to maintain a cache of 12 images, ready for users whenever they tap a cell for random generation. We have found it impossible under normal usage to tap quickly enough to exhaust the cache for random generation. This produces a feeling of having a varied set of interesting images *instantly* on hand. Moreover, the crossover and mutation of an image to produce eight offspring is also usually extremely fast, providing a fluid, near-instant exploration of the space.

Users can tap an image on the sheet to see it re-rendered at 2000×2000 pixels (as in the fourth screenshot of Fig. 1). This enables zooming in to see details, which can be fun. Re-rendering is achieved by scaling the particle positions, drawing 4× larger shapes and blurring with a 4× larger radius. On average, the larger rendering takes 4 times as long as the smaller one, on all devices. Hence the average time to generate larger images is 3.44, 1.48 and 1.32 s on the iPad, iPhone and iPad Pro respectively. Users tend not to mind this wait, but it does represent room for improvement, and we are currently working on a multi-threaded generation process for the larger images. As described in the next section, random generation in Art Done Quick is further guided by a hand-curated database of 5,000 genomes pre-selected by a machine vision system.

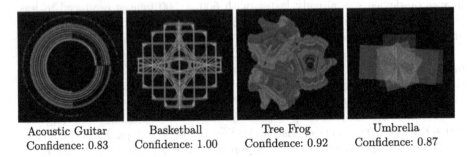

| Acoustic Guitar | Basketball | Tree Frog | Umbrella |
| Confidence: 0.83 | Confidence: 1.00 | Confidence: 0.92 | Confidence: 0.87 |

Fig. 4. Images which were strongly categorised (with confidence $\geq$0.8) by Resnet. (Color figure online)

4   Machine Vision Enhancements

Recently, neural machine vision systems have become highly effective at solving image classification problems, i.e., predicting aspects of the contents of a digital image. We experimented with three pre-trained image classifying neural models that are available for iOS developers, namely MobileNetV2 [23], Resnet50 [15] and SqueezeNet [14,17]. These have all been trained on the ImageNet archive of labelled images [8], and for a given 224×224 pixel image, they return a confidence score (between 0 and 1) for the prediction that the ground truth of the image is in each of 1,000 image categories. The highest-confidence category is normally taken as the prediction coming from the model.

We were motivated to consider using image classification technology through some initial experiments with Resnet. In particular, we found that it could reliably see something in an image that we couldn't initially see ourselves, but could identify on further inspection. That is, on reading Resnet's image classification, we were often encouraged to look again at an image and try to see the object/scene it predicted. Moreover, it was usually pleasurable to solve the initial recognition puzzle and rewarding when we convinced ourselves that we also saw the essence of the prediction in the image. To decrease the number of frustrating times where we couldn't fathom the Resnet predictions, we concentrated on images where it predicts a category with 0.8 confidence or higher. We found that the images with confidence predictions over this threshold almost always convey some aspect of the category predicted, but – as the images are abstract – this is usually not obvious at first glance. As our initial experiments proceeded, our trust in Resnet increased with usage, so that we were often encouraged to believe it more than our own eyes, and look harder to understand its prediction.

As good examples of the enjoyment of encountering such predictions, in Fig. 4, the first image was predicted with 0.83 confidence by Resnet to be an acoustic guitar. It took us a while to realise that the image captured the essence of only the sound hole in a guitar, not the entire instrument. On realising this, we noted that the image nicely portrays such a sound hole in shape and colour and possibly conveys the mood of a performance in a darkened venue. The second image captures aspects of the shape and texture of the basketball that Resnet predicted, but we can also clearly see aspects of a basketball *net*. The third image captures the colour (green skin, red eyes) and spindly limbs of the tree frog that Resnet predicts with 0.92 confidence. Producing a portrayal of a living creature in this way has echoes of Latham's Mutator images [28] and McCormack's flora and fauna images [20]. In these first three images, there is little chance that we would have seen the object/animal as predicted, but in each case, we took pleasure in agreeing with what Resnet saw. In the final image of Fig. 4, we did pre-identify an umbrella, and it was satisfying that Resnet agreed with us. In the context of a casual creator, we felt that – if used appropriately – such content prediction in abstract art images has the potential to be a pleasurable experience and a suitable addition to the Art Done Quick app.

4.1 Deriving a Machine Vision Setup

To derive a suitable vision model, we performed some experiments comparing Resnet, MobileNet and SqueezeNet, in the context of potentially adding functionality where Art Done Quick occasionally invents titles for images. For instance, we could imagine "Umbrella Rainbow" as an appropriate title for the fourth image in Fig. 4. The hope is that such titles would encourage users to see their creations through fresh eyes, in an enjoyable fashion, with notable *lightbulb moments* when visual puzzles are solved. To increase satisfaction, we needed to find a balance where the predicted image category is difficult to agree with at first, but difficult to disagree with after a more lengthy consideration.

In the context of machine vision, an adversarial example is an image which has been altered specifically to be misclassified by a particular vision model, while not changing the ground truth class of the image [26]. Normally, such examples are constructed with access to the model being attacked, but Liu et al. showed that adversarial examples can transfer from one machine vision model to another [19]. In our case, Art Done Quick's images are not constructed to look like they contain any particular content, so there is no ground truth. Hence while they are not adversarial examples in the strict sense, any image for which a machine vision model strongly predicts a category is a false positive for that model, so they may share some properties with respect to fooling a classifier. For our lightbulb-moment context, we are interested in images which are difficult to classify by people, at least temporarily. A similar phenomenon was studied in [10], which considered and falsified the assumption that adversarial examples for machines are not adversarial for people. In particular, the main finding was that if an adversarial example fools numerous machine classifiers, it will likely transfer to be adversarial for people, under conditions where there is limited time to make a decision. This leads to the hypothesis that the best types of images for lightbulb moments in our context will be those which are strongly categorised in the same way by multiple image classifiers.

To start investigating this potential, we have undertaken some analysis of the quality and quantity of predictions made by the three models, to test the feasibility of producing images which are false positives for multiple image classifiers. We first produced a confidence profile for the MobileNet, Resnet and SqueezeNet models, by generating 2,000 images randomly (as described above), and recording the highest confidence for any category that it saw, along with the number of different top-prediction categories it encountered during the session. The results are given in Table 1. We first note that the three models classified the images into only around 15–17% of the 1,000 ImageNet categories. For all three models, the average confidence is lower than 0.5, with SqueezeNet being the most confident at 0.4822 on average, Resnet next at 0.3751 and MobileNet at 0.2692. This low confidence is to be expected for abstract art images. We see that our informal confidence threshold of 0.8 for Resnet separates the top 12% of images from the rest. With this in mind, we likewise set 0.6 as a threshold for MobileNet (separating 11% of the images), and 0.9 for SqueezeNet (separating 12%). Any image categorised above the relevant threshold for a classifier can be said to be **strongly predicted for** by the model. We also say that an image is **weakly predicted for** by a model if it is not strongly predicted, but the confidence is above the relevant average confidence in Table 1, i.e., above 0.2692, 0.3751 and 0.4822 for MobileNet, Resnet and SqueezeNet respectively.

Table 1. Top category confidence profiles over 2,000 images for the vision models.

Neural model	Confidence bands										Av. highest confidence	Different categories
	0–.1	.1–.2	.2–.3	.3–.4	.4–.5	.5–.6	.6–.7	.7–.8	.8–.9	.9–1.0		
Mobile net	550	472	315	199	131	118	80	45	55	35	0.2692	156
	28%	24%	16%	10%	7%	6%	4%	2%	3%	2%		
Resnet	287	392	306	264	189	140	83	109	119	111	0.3751	150
	14%	20%	15%	13%	9%	7%	4%	5%	6%	6%		
Squeeze net	276	144	222	220	236	170	180	160	145	247	0.4822	168
	14%	7%	11%	11%	12%	8%	9%	8%	7%	12%		

Using these weak and strong boundaries, and recalling that our aim is for Art Done Quick to occasionally suggest a title for an image, we ran another experiment to evaluate the yields of numerous schemes for choosing which images to label. Each scheme has a **primary model** and 0, 1 or 2 **secondary models**. The top category of the primary model is the prediction taken for the scheme, which can be strong or weak, as above. We say that a secondary model **strongly supports** the primary prediction if the confidence it has for that prediction is over its particular strong prediction boundary, and likewise for **weak support**.

We label a scheme with upper and lower case letters from {M, R, S, m, r, s} as follows. The first letter of the scheme label denotes the primary model as (M)obileNet, (R)esnet or (S)queezeNet, with uppercase representing a strong prediction and lower case a weak prediction required for the scheme to select an image to label with the primary prediction. The (optional) second and third letters represent the secondary models and whether they need to give strong (uppercase) or weak (lowercase) support in order for the scheme to choose to perform the labelling. As an example, Mrs denotes the scheme where MobileNet is the primary model requiring a strong prediction, weakly supported by Resnet and SqueezeNet secondary models.

We generated another 2,000 images randomly, and recorded the number of images that each of numerous schemes chose to label, with the results in Table 2. We first see that, for the kinds of image that Art Done Quick makes, there is very little agreement in the predictions from the three machine vision systems. In fact, of 2,000 images generated, only 2 (0.1%) were strongly predicted to contain the same content by all three models. Of the schemes which used a strong prediction weakly supported, Rm had the highest yield, with 82 images (4.1%). Of those using a weak prediction weakly supported, rm and mr yielded the most with 184 images (9.2%). If high yield and some level of quality is required, the "M, R or S" scheme – where any model strongly predicted for an image – yielded 588 images (29.4%), but the proportion of these supported by a secondary model is very low. Nearly three quarters of the images (1,467 of 2,000) were weakly predicted for by at least one model (the "m, r or s" scheme in Table 2).

Table 2. Success profile of schemes to decide on labelling images with machine vision categories, over 2,000 images.

Scheme	M	R	S	m	r	s	M, R or S	m, r or s	
Chosen	201	222	274	734	824	953	588	1467	
Categories	33	40	53	70	96	115	136	642	

Scheme	MR	MS	RM	RS	SM	SR	Mr	Ms	Rm
Chosen	34	9	34	9	9	9	75	17	82
Categories	12	5	12	6	5	6	24	9	20

Scheme	Rs	Sm	Sr	mr	ms	rm	rs	sm	sr
Chosen	21	21	24	185	48	184	59	51	60
Categories	12	10	10	31	18	31	20	19	20

Scheme	MRS	RMS	SMR	Mrs	Rms	Smr	mrs	rms	smr
Chosen	2	2	2	13	11	9	20	22	22
Categories	2	2	2	9	9	6	13	13	13

4.2 Deploying the Vision Model in Art Done Quick

We are currently trialling the Rm scheme in Art Done Quick, as this chooses to label roughly 1 in 25 random images, which seems appropriate. It has the benefit that images must pass the 0.8 Resnet confidence threshold which we have learned to trust, and an added benefit of weak support from the classification of MobileNet. One drawback to this scheme, however, is the small number of categories that it strongly classifies images into – with only 20 different categories used in classifying 2,000 images (see Table 2). Note that Resnet alone only used 40 different categories out of 1,000 (4%). In practice, this means that users will very likely lose interest in the labelling, as they only see a small number of labels used repetitively. While this could be mitigated somewhat with natural language processing (using WordNet, for example, as the categories are mapped from there), it would be better to increase the variety of images generated, to include images strongly classified into a wider set of categories.

To this end, we have performed additional experimentation to produce images that increase the number of categories Resnet uses. We implemented a hill-climbing approach where images are first generated randomly and categorised by Resnet. Any which have a strong confidence (≥ 0.8) are recorded in a database, along with the category Resnet predicted. If an image, I, is not strongly predicted for, but is weakly classified into category C by Resnet (i.e., with confidence above the average of 0.3751 from Table 1), then if C has not been seen before in the database, hill climbing starts. This entails producing mutations of I until either a certain number, H, are tried, or a mutation is found that Resnet strongly predicts to be in C, which is recorded in the database. Note that each mutation is performed on the image with the highest confidence for C seen so far, which may not be the original. During hill climbing, any image with Resnet confidence over 0.8 for some category other than C is recorded in the database.

Table 3. Hill climbing sessions checking 5,000 images for strong classification by Resnet. The session setups differed by the amount of hill climbing allowed, H.

Steps (H)	0	10	20	30	40	50	60	70	80	90	100
Categories	56	70	73	69	63	68	65	75	63	70	61
Images	508	483	513	469	449	435	390	419	399	432	445
Success/fail	–	37/258	31/148	31/110	31/86	29/71	33/57	44/42	27/46	46/33	31/33
Av. length	–	5.91	9.84	12.70	15.20	20.64	26.156	24.07	22.25	28.90	37.47

A higher value for H means that more hill climbing attempts will be success-ful, as they are given more opportunity to climb higher. However, a bigger H value also means that fewer hill climbing attempts will be started. To investigate this trade-off, we undertook 11 sessions where Art Done Quick generated 5,000 images, varying H from 0 to 100 in steps of 10. We recorded how many different categories of images were predicted for by Resnet, how many images in total were above the strong threshold, how many times the hill climbing effort was successful and failed, and finally the average length of the successful hill climbs. The results are given in Table 3. We note that without hill climbing ($H = 0$ in Table 3), random generation alone achieves fewer categories than with hill climbing. We found that random generation only achieves categories which are *easy*, in the sense of the limited styles of output from Art Done Quick regularly looking like images in those categories. These easy categories soon run out, and the hill climbing becomes necessary. We also note that the average length of the successful hill climbs did not plateau, so the search was taking advantage of longer hill climbing opportunities, which we also see in the balancing of success/failure in hill climbing attempts as H increases in Table 3.

Taking these results into account, we introduced a sliding scale approach whereby the search starts with $H = 20$, and every time images in 50 more new categories are seen, H is increased by 10. We found that this was effective in finding varied images that Resnet classified strongly into a large number of categories. We ran it in five separate threads on an iPad Pro, and were able to produce and test 500,000 images over a series of sessions, with the value of H and the database shared between threads and sessions. This led to 71,539 strongly categorised images, placed into 408 different categories. The categories with the most images are jellyfish (9838 images), sea urchin (6986) and oscilloscope (6304). In the database, there are 13 categories with 1,000 or more images in and 61 categories with 100 or more. In contrast, there are 120 categories with a single image, and 223 categories with five or fewer images in. We plan to increase the number of categories and the number of images per category by experimenting with evolutionary searches involving both mutation and crossover.

To take advantage of the database for Art Done Quick, we hand-curated example genomes for images from each of the 408 categories, to keep as seeds for the random generation, discussed below. The fact that the images cover so many ImageNet categories means that the imagery is quite diverse already, but we further selected images not just for quality (subjectively), but also for diversity. To do this, for the 347 categories with fewer than 100 images, we looked at each of the images, totalling 3,878 images. For the 61 categories with 100+ images, we used k-means clustering with k set to 50 to narrow the choices into visually distinctive clusters, with a representative image chosen as being closest to the centroid of each cluster. This resulted in another sample of 3,050 images which we hand-curated also. We found the quality and diversity of the images to be very high, and kept 5,000 images in total. The values used to analyse images for the clustering came from the usage of a *headless* version of Resnet, where the final layer of the artificial neural network is removed, and the model outputs 2,048 floating point numbers which capture visual aspects of the images input. Such a headless technique is often used in image retrieval, as it was in [25].

When users tap on an empty cell in Art Done Quick, they are given an image generated from one of the 5,000 database genomes with 0.25 probability, from a mutated version of a database genome with 0.25 probability, from a genome obtained by crossing over two of the 5,000 genomes with 0.25 probability, and randomly generated with 0.25 probability. We have found that this has

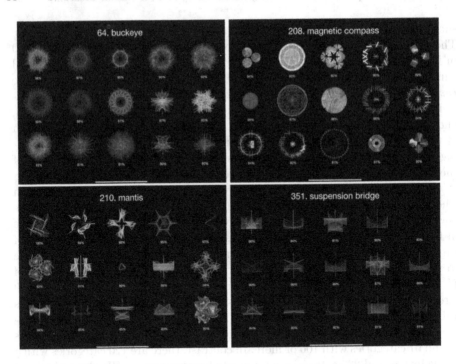

Fig. 5. Four categories of starting point images (iPad version, landscape).

greatly increased the variety of labels that setup Rm provides users, and it does so for around 1 in every 10 images, a frequency which we will test against the satisfaction levels of users. The database has also enabled functionality where users can choose images as entry points to the casual creation. Currently, images are arranged by the ImageNet categories which users can browse, as depicted in Fig. 5, but we plan to use more artistic categories, which will also address the fact that some Resnet categories have few genomes in the database. In a final enhancement of Art Done Quick using machine vision, we used the clustering approach mentioned above to group together sets of images, on-demand when the user wishes to re-arrange the image sheet depicted in Fig. 1. This can often tidy up the sheet, enable users to find images or suitable pairs for crossover more easily, and/or heighten an enjoyable feeling of collecting and curating images.

5   Related Work

Casual creators are creativity support tools [24] designed to maximise fun in their usage, as per the HCI subfield of *funology* [2]. However, creativity support is usually aimed at improving the productivity of already creative people, often creative practitioners, rather than members of the public, so best practice advice often doesn't transfer, and has to be tensioned against best practice in funology.

There is also much overlap with the study of ludic experiences and games, particularly in areas such as human curiosity [21], and mixed-initiative co-creativity, where creative responsibility is shared between human and computer [30].

A new wave of *GAN artists* are using neural models in art practice, via techniques such as generative adversarial networks [13], style transfer [11] and style invention [9]. For instance, in [1], the authors used adversarial examples to co-evolve convolutional neural networks as a simulated artist and critic. In [27], Tanjil and Ross used a deep convolutional network for image classification as a fitness function to evolve abstract images which resemble certain objects, like a peacock. Artists such as Tom White have also used neural image classification systems in their work [29], in this case with a *perception engine* which constructs images with small numbers of shapes and lines resembling brush strokes. The images are optimised against a neural classification model, with successful ones being printed in a process similar to screen printing. Colton et al. similarly used Resnet in [5], with their HR3 system producing programs which placed geometric shapes onto a canvas, so that a plotter could produce physical realisations. Tens of thousands of images were passed through Resnet and ones with a strong classification were hand-picked for plotting and exhibited in a pop-up show.

6   Conclusions and Future Work

While the work presented here is grounded in the development of the Art Done Quick app, there are a number of general points to take away. Firstly, we should not ignore the fact that using generative visual art systems can be an enjoyable process, over and above the value of the images produced. Not every generative art system will find usage in fine art practice, but each has the potential to provide a fun user experience, and this can drive design decisions as much as image quality or professional usage. We have motivated and described above a fun-first design methodology, and we will be considering further casual creator design patterns from [7] for Art Done Quick. In particular, there is no managed sense of achievement in using the app, and we will consider ways in which gamification can be used to add this (always asking whether the addition will increase enjoyment). Also, Compton et al. point out that casual creators are often used in a social way, with users sharing, modifying, curating and commenting on the creations of others as well as their own. Again, we will investigate fun ways to achieve such sharing and community building via Art Done Quick.

Secondly, neural machine vision can bring a number of benefits to visual art generation systems, in terms of improved search and interesting evaluations. The usage of image classifiers trained on real-world images, but deployed in artistic contexts, is still in its infancy: we can expect to see much more varied and interesting uses of this technology in the arts. We will be experimenting with using additional machine vision technologies within Art Done Quick, including using image segmentation and depth perception models, as well as heatmap visualisations to find salient regions of an image. It is also possible to update a pre-trained neural model on-device to adapt to user preferences, and we plan to use this to learn aesthetic preferences for individual users.

Clearly, we will only know with certainty whether users have fun while playing with the Art Done Quick app after we have undertaken substantial user testing and analysed the results. However, we feel that we have given the app a good chance at being successful in this respect by (i) adhering to many design patterns identified in [7] as leading to enjoyable casual creator usage (ii) drawing on the success of existing published casual creators and (iii) experimenting to reduce frustrations inherent in evolutionary art generation. Compton et al. talk about casual creator apps having seemingly *magical* elements, which often derive from their generative aspects. We hypothesise that there will be three such elements in Art Done Quick: (a) the instant and endless generation of varied and valued images at users' fingertips (b) the carrying over of image edits into offspring, and (c) the occasional moment when the machine vision scheme poses an intriguing visual puzzle which is solved to much satisfaction. In stage two of the app development, we will test the magic of these features, and more generally assess which aspects of the app usage are enjoyable and which are not.

We plan to use Art Done Quick as a research platform on which to test psychological theories of flow, engagement and immersion of users while undertaking casual creation. We will also employ it in a Computational Creativity context, by adding a second layer comprising an AI system that uses the app in the same way that people do. In a piece entitled "Can you see what I can see?", the secondary AI system will be aiming to make images such as those in Fig. 4, to divide opinion amongst people about what can be seen in abstract imagery. The piece will use the evolutionary engine and the image editing features of Art Done Quick, and has been selected for installation in two exhibitions in 2020. One aim for the piece is to highlight the difference between AI systems that make decisions during the creative process, and those that do not.

Acknowledgements. We would like to thank the anonymous reviewers for their comments. Many thanks also to the SensiLab colleagues and IGGI students who gave such detailed, insightful and supportive feedback on the Art Done Quick app.

References

1. Blair, A.: Adversarial evolution and deep learning – how does an artist play with our visual system? In: Ekárt, A., Liapis, A., Castro Pena, M.L. (eds.) EvoMUSART 2019. LNCS, vol. 11453, pp. 18–34. Springer, Cham (2019). https://doi.org/10.1007/978-3-030-16667-0_2

2. Blythe, M., Monk, A.: Funology 2: From Usability to Enjoyment. Springer, Heidelberg (2018). https://doi.org/10.1007/978-3-319-68213-6

3. Colton, S., Cook, M., Raad, A.: Ludic considerations of tablet-based evo-art. In: Di Chio, C., et al. (eds.) EvoApplications 2011. LNCS, vol. 6625, pp. 223–233. Springer, Heidelberg (2011). https://doi.org/10.1007/978-3-642-20520-0_23

4. Colton, S.: Evolving a library of artistic scene descriptors. In: Machado, P., Romero, J., Carballal, A. (eds.) EvoMUSART 2012. LNCS, vol. 7247, pp. 35–47. Springer, Heidelberg (2012). https://doi.org/10.1007/978-3-642-29142-5_4

5. Colton, S., Pease, A., Cook, M., Chen, C.: The HR3 system for automated code generation in creative settings. In: Proceedings of the 10th ICCC (2019)

6. Compton, K.: Casual creators: AI supported creativity for casual users. Ph.D. thesis, University of California, Santa Cruz (2019)
7. Compton, K., Mateas, M.: Casual creators. In: Proceedings of the 6th International Conference on Computational Creativity (2015)
8. Deng, J., Dong, W., Socher, R., Li, L.-J., Li, K., Li, F.-F.: ImageNet: a large-scale hierarchical image database. In: Proceedings of IEEE CVPR (2009)
9. Elgammal, A., Liu, B., Elhoseiny, M., Mazzone, M.: CAN: Creative adversarial networks. In: Proceedings of the 8th ICCC (2017)
10. Elsayed, G., et al.: Adversarial examples that fool both computer vision and time-limited humans. In: Proceedings of 32nd International Conference on Neural Information Processing Systems (2018)
11. Gatys, L., Ecker, A., Bethge, M.: A neural algorithm of artistic style. In: Proceedings of IEEE Computer Vision and Pattern Recognition Conference (2016)
12. Gaudl, S., et al.: Rapid game jams with fluidic games. Entertain. Comput. **27**, 1–9 (2018)
13. Goodfellow, I., et al.: Generative adversarial networks. In: Proceedings of NIPS (2014)
14. Han, S., Mao, H., Dally, W.: Deep compression: compressing deep neural networks with pruning, trained quantization and Huffman coding. In: Proceedings of ICLR (2015)
15. He, K., Zhang, X., Ren, S., Sun, J.: Deep residual learning for image recognition. In: Proceedings of IEEE Computer Vision and Pattern Recognition Conference (2016)
16. Hull, M., Colton, S.: Towards a general framework for program generation in creative domains. In: Proceedings of 4th International Joint Workshop on computational Creativity (2007)
17. Iandola, F., Han, S., Moskewicz, W., Ashraf, K., Dally, W., Keutzer, K.: SqueezeNet: AlexNet-level accuracy with 50x fewer parameters. arXiv:1602.07360 (2016)
18. Iyengar, S., Lepper, M.: When choice is demotivating: can one desire too much of a good thing? J. Pers. Soc. Psychol. **79**, 995 (2000)
19. Liu, Y., Chen, X., Liu, C., Song, D.: Delving into transferable adversarial examples and black-box attacks. In: Proceedings of International Conference on Learning Representations (2016)
20. McCormack, J.: Aesthetic evolution of L-systems revisited. In: Raidl, G., et al. (eds.) EvoWorkshops 2004. LNCS, vol. 3005, pp. 477–488. Springer, Heidelberg (2004). https://doi.org/10.1007/978-3-540-24653-4_49
21. Nelson, M., Gaudl, S., Colton, S., Deterding, S.: Curious users of casual creators. In: Proceedings of the 13th Conference on the Foundations of Digital Games (2018)
22. Romero, J., Machado, P.: The Art of Artificial Evolution: A Handbook on Evolutionary Art and Music. Springer, Heidelberg (2008). https://doi.org/10.1007/978-3-540-72877-1
23. Sandler, M., Howard, A., Zhu, M., Zhmoginov, M., Chen, L.-C.: MobileNetV2: inverted residuals and linear bottlenecks. In: Proceedings of IEEE CVPR (2018)
24. Shneiderman, B.: Creativity support tools: accelerating discovery and innovation. Commun. ACM **50**(12), 20–32 (2007)
25. Singh, D., Rajcic, N., Colton, S., McCormack, J.: Camera obscurer: generative art for design inspiration. In: Ekárt, A., Liapis, A., Castro Pena, M.L. (eds.) EvoMUSART 2019. LNCS, vol. 11453, pp. 51–68. Springer, Cham (2019). https://doi.org/10.1007/978-3-030-16667-0_4

26. Szegedy, C., et al.: Intriguing properties of neural networks. In: Proceedings of ICLR (2013)
27. Tanjil, F., Ross, B.J.: Deep learning concepts for evolutionary art. In: Ekárt, A., Liapis, A., Castro Pena, M.L. (eds.) EvoMUSART 2019. LNCS, vol. 11453, pp. 1–17. Springer, Cham (2019). https://doi.org/10.1007/978-3-030-16667-0_1
28. Todd, S., Latham, W.: Evolutionary Art and Computers. Academic Press, Cambridge (1992)
29. White, T.: Perception engines (2018). https://medium.com/artists-and-machine-intelligence/perception-engines-8a46bc598d57
30. Yannakakis, G., Liapis, A., Alexopoulos, C.: Mixed-initiative cocreativity. In: Proceedings of the 9th Conference on the Foundations of Digital Games (2014)

Comparing Fuzzy Rule Based Approaches for Music Genre Classification

Frederik Heerde[1,2], Igor Vatolkin[1(✉)], and Günter Rudolph[1]

[1] Department of Computer Science, TU Dortmund, Dortmund, Germany
{Frederik.Heerde,Igor.Vatolkin,Guenter.Rudolph}@tu-dortmund.de
[2] QuinScape GmbH, Dortmund, Germany

Abstract. Most of the studies on music genre classification are focused on classification quality only. However, listeners and musicologists would favor comprehensible models, which describe semantic properties of genres like instrument or chord statistics, instead of complex black-box transforms of signal features either manually engineered or learned by neural networks. Fuzzy rules – until now not a widely applied method in music classification – offer the advantage of understandability for end users, in particular in combination with carefully designed semantic features. In this work, we tune and compare three approaches which operate on fuzzy rules: a complete search of primitive rules, an evolutionary approach, and fuzzy pattern trees. Additionally, we include random forest classifier as a baseline. The experiments were conducted on an artist-filtered subset of the 1517-Artists database, for which 245 semantic properties describing instruments, moods, singing style, melody, harmony, influence on listener, and effects were extracted to train the classification models.

Keywords: Fuzzy rule bases · Music genre recognition · Semantic features

1 Introduction

Fuzzy sets, first introduced in [15], extend classical sets by allowing elements to be their members to a certain degree μ. This makes it possible to express concepts mathematically that have ambiguous transitions. Furthermore, they serve as a basis for fuzzy logic, whose fuzzy statements, such as "Tempo is *high*", have a degree of truth between 1 (true) and 0 (false). Fuzzy statements can be negated and combined with fuzzy operators. In this work, we use NOT, AND, OR, and AVG, a special case of an operator proposed in [6] that calculates the average degree of truth of two fuzzy statements. We examine three fuzzy rule based classification approaches and apply them to music genre recognition. During the training stage, they build a set of fuzzy rules for every genre. Together, these sets form the rule base. The rules consist of a fuzzy statement as premise and a genre as conclusion, e.g., "IF Mood Reflective is *very high* THEN Jazz". A prediction

J. Romero et al. (Eds.): EvoMUSART 2020, LNCS 12103, pp. 35–48, 2020.
https://doi.org/10.1007/978-3-030-43859-3_3

is made by determining the genre for which the premises of its rules yield the highest degree of truth on average.

Because predictions based on fuzzy rules can be expressed in natural language, the classification models can be very comprehensible, as long as the used features are understandable as well. This is an advantage over other classifiers with complex black-box models and makes fuzzy rules favorable for end users. On the other side, these features cannot always be robustly extracted from complex polyphonic audio signals. Still, it was shown in [9] that high-level features are a good basis for classification. Furthermore, rule bases can be easily modified on request by music experts.

We compare three fuzzy rule based classification approaches. The first approach from [13] is a complete search that calculates a relevance measure for every possible primitive rule and uses the most relevant rules for classification. The second approach, called fuzzy pattern trees [11], constructs one fuzzy rule per genre iteratively in a tree like manner. The third approach is a modification from an evolutionary algorithm proposed in [8]. It generates random rule bases and optimizes them via mutation and recombination. In our experiments, we compare these methods with a random forest classifier as a baseline. To ensure interpretability, we classify genres only based on semantic mid-level features like instruments, harmonic properties, predicted moods, etc.

The remainder of this paper is organized as follows. First, Sect. 2 outlines the work related to ours. In Sect. 3, we explain the operating principles of the investigated algorithms. The experiments are presented in Sect. 4. Lastly, we draw a conclusion in Sect. 5.

2   Related Work

When using primitive fuzzy rules in the rule base, a complete search can be executed, shown in [13]. They established a measure for the relevance of one rule. All possible primitive rules were sorted according to this measure in an descending order. When classifying, they used the m most relevant primitive rules, where the optimal m needed to be determined. This is not feasible for rule bases with complex rules, though, as the number of possibilities increase drastically.

An interesting alternative are fuzzy pattern trees, that were originally proposed in [6] and further explored in [10,11]. By constructing the rules incrementally and only keeping track of the most promising ones, the number of possible rules to evaluate was greatly reduced as opposed to a complete search.

Furthermore, there are numerous evolutionary approaches to build fuzzy rule bases. They differ in the representation of the algorithm's solution, or individual. For example, in both [5] and [1], an evolutionary algorithm was proposed, in which individuals represent single rules. However, their approaches differ in the way they form the rule base from the result of the algorithm. [5] followed an iterative approach, in which an evolutionary algorithm generates one rule at a time, which is added to the rule base. After each iteration, a boosting algorithm gives more weight to the training examples that the current rule base misclassifies. This affects the fitness function used by the evolutionary algorithm to enforce a rule that covers the misclassified training examples better. In contrast, the algorithm in [1] used the best population as a rule base. To prevent homogeneous rule bases, they incorporated a measure for diversity in their fitness function. Afterwards, they simplified the rule bases by removing redundant rules. An individual can also be interpreted as the whole rule base, as seen in [8]. This is also the approach we chose as a basis for ours.

The fuzzy sets that are used to fuzzify feature values can be optimized after the model was trained, as done with a genetic algorithm in [3,4].

Apart from fuzzy rule based approaches, there exist also fuzzy classification methods, in which models assign a degree of membership to each class instead of making a crisp prediction. Here, we give a few examples related to music classification. In [14], a fuzzy k-NN classifier and a fuzzy Nearest-Mean classifier were applied for the fuzzy classification of music to emotions. In [7], a modification of the fuzzy C-Means clustering algorithm was used to group pieces of music by genre in the feature space. Calculating the distance to each cluster centroid results in the wanted degrees of membership to each class.

3 Methodology

3.1 Random Forest

As a baseline, we included random forest (RF) in our experiments. This classification algorithm was proposed in [2] and is an ensemble of small decision trees which are trained with a randomly selected subset of all features. RF has several advantages: It is fast, robust to overfitting, and has only a few parameters to setup. On the other side, the models are usually less interpretable, because the good robustness is guaranteed only if a sufficiently large number of trees (e.g., 100) are created.

3.2 Complete Search for Primitive Rules

In [13], a method was proposed that uses a set of primitive fuzzy rules for music genre classification. Here, a primitive fuzzy rule is defined as IF <feature X> is <*linguistic term* T> THEN <genre C>. A measure was established to estimate the relevance of a single primitive rule for a genre in an annotated training set:

$$R(C, X, T) = P(X \text{ is } T \mid C) \cdot (1 - P(X \text{ is } T)) \qquad (1)$$

Feature values were assigned to linguistic terms like *low, moderate*, or *high*. Firstly, the method generated every possible primitive rule for one genre. Secondly, the relevance was calculated for every rule. Finally, the rules were sorted by relevance in descending order. This was done for every genre using a One-vs-All scheme, i.e. a binary classification of music pieces as either exclusively belonging to a given genre or to other genres.

To classify a piece of music, the m most relevant rules were estimated for every genre. Then, this piece is assigned to a genre with the highest average relevance of these rules.

3.3 Fuzzy Pattern Trees

Fuzzy Pattern Trees (FPTs) were introduced in [6]. Essentially, a FPT is a fuzzy rule, whose premise is interpreted as a tree, in which fuzzy operators serve as inner nodes and fuzzy statements as leafs. A class can be represented by one FPT only.

In [10], a top-down deterministic algorithm for constructing fuzzy pattern trees was introduced. They also presented other stopping criteria in [11], as well as a faster, yet nondeterministic version of the algorithm. Here, we use the deterministic method with the stopping criteria from [11].

Basically, the algorithm keeps track of one rule $\mathcal{M}^*$ that is extended incrementally. It initializes $\mathcal{M}^*$ by selecting the fuzzy statement with the smallest error rate on a training set $\mathcal{T}$. In each iteration, $\mathcal{M}^*$ gets extended. A tree gets extended by replacing one leaf, i.e. a fuzzy statement, with a fuzzy operator that has a new statement and the old leaf as a child. From all possible extensions that do not result in a tree exceeding the maximum depth, the one with the smallest error is selected. The algorithm terminates if every branch of $\mathcal{M}^*$ has reached the maximum depth or the error after the extension was higher than $(1 + \gamma)$ times the error before the extension, where γ is a parameter that specifies by how much the new $\mathcal{M}^*$ is allowed to be worse than the last one.

As the error rate, the root mean squared error was used, just like in [11]. Consider a set $\mathcal{T}$ of tuples (x, y) that are constituted of a feature vector x and the membership y of this vector to a class, expressed as 0 or 1. For a fuzzy pattern tree $\mathcal{M}$, the root mean squared error can be calculated as shown in Eq. 2. $\mathcal{M}(x)$ denotes the output of the model when given the feature vector x, also as 0 or 1.

$$m_{\mathcal{T}}(\mathcal{M}) = \sqrt{\frac{1}{N} \sum_{(x,y) \in \mathcal{T}} (y - \mathcal{M}(x))^2} \qquad (2)$$

3.4 Evolutionary Approach

In [8], an evolutionary algorithm was developed to find a rule base for a fuzzy system. Its rules conclude the membership to another fuzzy set, e.g., "IF feature$_1$ is *high* THEN feature$_2$ is *low*". One individual represents a rule base.

We adjusted this approach to fit the purpose of music genre classification. Now, rules for classification will be used, like "IF feature$_1$ is *high* THEN class$_3$". First experiments have shown that it is beneficial to train the rule base for one genre at a time. After the algorithm terminates for each genre separately, the best genre-specific rule bases are joined to form a rule base that covers every genre. Consequently, one individual represents a set of rules for one genre.

To evaluate the quality of an individual $\mathcal{M}$, the root mean squared error is calculated and subtracted from 1, as shown in Eq. 3. In this case, the individual $\mathcal{M}$ is a set of rules and the degree of truth of every of its rules $\mathcal{M} \in \mathcal{M}$ can be calculated separately by inputting a feature vector x: $\mathcal{M}(x)$.

$$f_T(\mathcal{M}) = 1 - \sqrt{\frac{1}{N} \sum_{(x,y) \in T} \left(y - \frac{1}{|\mathcal{M}|} \sum_{\mathcal{M} \in \mathcal{M}} (\mathcal{M}(x)) \right)^2} \tag{3}$$

Procedure. We consider four parameters for this algorithm: the population size $\#_{pop}$, the initial number of rules per individual $\#_{rule}$, the number of generations gen_{max}, and the last parameter to control the complexity of random premises d_{rule}. This procedure starts for each genre separately.

First, the population gets initialized with $\#_{pop}$ random individuals. One individual consists of $\#_{rule}$ rules with randomly generated premises for the current genre. In each generation, the algorithm iterates over every individual in the current population. The individual will be either recombined or mutated. Thus, only one probability is needed to decide between mutation and recombination. For recombination, the algorithm selects a second individual via the roulette method, where the probability for an individual to be picked is proportional to its fitness. The new population will be selected from the old population and all newly generated individuals. Following [8], the last step in each generation is to linearly reduce the recombination probability in such a way that it equals 20% for the last generation. Consequently, the mutation probability reaches 80% at last.

Generating Random Premises. The process of generating random premises from [8] interprets premises as trees just like the FPTs described in Subsect. 3.3. It constructs a random premise top-down. With a fixed probability of d_{rule}, a random fuzzy operator is inserted into the root of the tree. Otherwise, a random fuzzy statement is composed and inserted. If a fuzzy operator was chosen, the same procedure begins for both children that are connected with the operator. Having a fixed probability for every layer of the tree has the disadvantage of the tree's depth being unbound. If d_{rule} is too low, most of the generated premises will be primitive. If it is too high, the emerged trees will be too huge and less interpretable.

Since we wanted to bound the trees' depth, we decided to change this aspect. Instead of having a fixed probability for inserts in every layer, it is reduced by 25% per layer. d_{rule} defines the probability on the top layer. Thus, the depths of the trees are limited since the probability reaches zero at some point. Other reduction schedules are also possible but were not part of our experiments, e.g. multiplying the probability with a number <1 per layer.

For example, when using $d_{rule} = 0.5$, 50% of the generated premises are primitive on average while the other half contain at least one operator. In these cases, the children of the root will be operators with the reduced probability of 25%.

Figure 1 shows an example for a randomly generated tree and the probabilities to select a fuzzy operator for each layer.

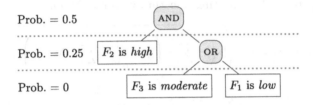

Prob. = 0.5

Prob. = 0.25 F_2 is *high*

Prob. = 0 F_3 is *moderate* F_1 is *low*

Fig. 1. Example for a randomly generated premise with $d_{rule} = 0.5$.

Mutation and Recombination. Both the mutation and recombination operator are implemented after [8].

Mutation consists of three different actions that are applied to one individual:

- Delete mutation: Deletes a random rule from the set.
- Insert mutation: Inserts a new randomly generated rule into the set.
- Rule mutation: Selects a rule from the set randomly and replaces a randomly selected subtree with a new randomly generated one.

Recombination is used to mix two individuals. Firstly, a random matching of the rules from the first individual to the rules of the second individual is calculated. For this purpose, the rules of each individual are permuted randomly. In the newly acquired order, pairs of rules are built by picking the rules from the individuals that share the same position. If one individual has more rules than the other, the remaining rules are matched with randomly chosen rules of the other individual. Thereby, every rule is part of at least one pair. Secondly, the rules of each pair are interpreted as a tree to perform a subtree crossover. Lastly, all mixed pairs are divided to form two set of rules again. These sets are the children of the recombination. Note that they have the same number of rules.

4   Experiments

4.1   Setup

For the classification of music pieces into genres, we use the publicly available 1517-Artists dataset[1], from which we selected a subset of 543 pieces of music that were assigned to six genres with the labels Classical, Electronic_and_Dance, Hip-Hop, Jazz, Rock_and_Pop, and R_and_B_and_Soul. The selection of the genres is equivalent to the one from [13]. In this subset, each artist is represented by exactly one track, to reduce the danger of learning the peculiarities of artists instead of genres.

---

[1] http://www.seyerlehner.info. Last accessed on 08.11.2019.

In accordance with [12, pp. 138–142], we have extracted 245 semantically understandable features from the audio recordings. An overview of feature groups is provided in Fig. 2.

Group	Feature
Share of an instrument	guitar, piano, wind, strings, drums
Mood	aggressive, energetic, sentimental, stylish, reflective, confident, earnest, partycelebratory
Singing style	clear, voice medium, rough, solo unison, solo woman, solo man, solo polyphonic
Melodic range	> octave, ≤ octave, linearly, volatile
Harmony	major, minor
Influence on listener	level of activation high
Effects	distortion

Fig. 2. Overview of all extracted features.

Then, we stored feature values for classification time windows of 5 s with 2.5 s overlap. In order to use the same fuzzy sets for every single feature, the feature values were normalized.

The data set was divided into five partitions for a 5-fold cross-validation, so that 80% of tracks were used for training and 20% for validation in each of the 5 repetitions.

As a basis for our fuzzy approaches, we first need to define the fuzzy sets that are used for the fuzzification of the feature values. Following [13], we used the same five linguistic terms and membership functions. They can be seen in Fig. 3.

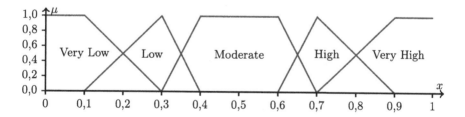

Fig. 3. Assignment of feature values to linguistic terms, originally from [13].

As described before, we obtained multiple feature vectors from each piece of music which will be used to train and validate the different approaches. By implication, the resulting models classify only an individual classification window of 5 s length represented by its feature vector. To make a prediction for the whole piece, we output the genre which was predicted most often among all classification windows of the same piece.

As error rate, we use the balanced relative error, written down in Eq. 4, where $m_{TP,c}$ and $m_{FN,c}$ are the numbers of true positives and false negatives for a class c and $|C| = 6$ is the overall number of predicted genres.

$$m_{BRE} = \frac{1}{|C|} \sum_{c \in C} \left(\frac{m_{FN,c}}{m_{TP,c} + m_{FN,c}} \right) \tag{4}$$

The evolutionary algorithm and the random forest were trained 15 times for each fold because of the nondeterministic behaviour of these methods. The complete search and fuzzy pattern trees are deterministic in the way we implemented them and hence, do not need repetition.

4.2 Results

Random Forest. We varied the number of trees between 100, 200, and 400. Figure 4 shows the results of the evaluation. The errors are averaged across 15 statistical repetitions and 5 folds ($15 \cdot 5 = 75$ values in total). Bold text indicates the smallest error. Note, that the expected error of a completely random classifier would be equal to 0.83. As it can be observed, the error rates differ only slightly, and using 400 trees yielded the best results. In succeeding graphics, we use these error rates as a baseline to compare with.

	Number of trees		
	100	200	400
Classical	**.0972**	**.0972**	**.0972**
Electronic_and_Dance	.1868	.1861	**.1825**
Hip-Hop	.4379	.4399	**.4357**
Jazz	.4770	.4763	**.4710**
R_and_B_and_Soul	.5272	.5182	**.5158**
Rock_and_Pop	.2714	.2703	**.2685**
Average	.3329	.3313	**.3285**

Fig. 4. Evaluation of random forest with 100, 200, and 400 trees.

Complete Search for Primitive Rules. Figure 5 plots the results of the complete search for primitive rules. Since the method is deterministic, it was executed only once for every fold. The error rate highly depends on m, the number of most relevant rules that are used for classification. When m is set to a low value, too few rules are incorporated per genre to accurately represent it, apparently. However, when a very high m such as 300 is used, not only the interpretability of classification models suffers, but also many less relevant rules contribute to the model. They are equally weighted as the most relevant rules and can therefore confuse the classifier.

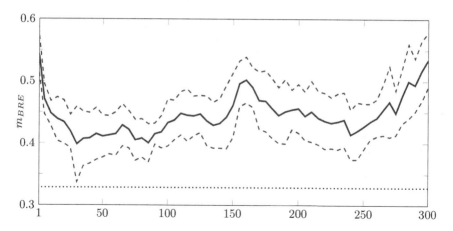

Fig. 5. Evaluation of the complete search for primitive rules when using the m most relevant rules for classification. The step size for m was 5. The solid line represents the mean validation error while the dashed lines indicate the standard deviation added or subtracted from it. The horizontal dotted line shows the validation error of the best RF model for comparison.

The algorithms were tuned on the training sets. We obtained the lowest training error and the lowest validation error of 39.77% for the same $m = 30$, meaning that this setting is robust enough for different data.

The 3 most relevant rules for the genre Jazz are shown in Fig. 6. Here, the feature "Level of Activation High" indicates how likely a music piece animates the listener. Please note that such semantic descriptors are often subjective by their definition, with regard to expert annotations and to music tracks selected to train a classifier for the prediction of a descriptor. They even may be confusing for some classification tasks like identification of specific Jazz subgenres. Just like in this example, it was generally noticeable that most relevant rules referenced either *"very low"* or *"very high"*. The fuzzy sets *"low"*, *"moderate"*, and *"high"* appeared in more irrelevant rules.

IF Level of Activation High is *very low* THEN Jazz
IF Mood Reflective is *very high* THEN Jazz
IF Mood Energetic is *very low* THEN Jazz

Fig. 6. The 3 most relevant rules for the training set of the genre Jazz.

Fuzzy Pattern Trees. The approach for the FPTs featured only one parameter d_{max} to determine the maximum depth of the trees. We tried out $d_{max} \in \{1, 2, 3, 4, 5\}$, where $d_{max} = 1$ generates primitive FPTs, that consist of one fuzzy statement only. Higher values for d_{max} were omitted because of too long training time and inferior interpretability. Note, that the time was reduced drastically in

[11] using various heuristics that rely on chance and thus, make the algorithm nondeterministic. We opted for the deterministic approach to ease the implementation and evaluation. For future work, the nondeterministic version might be a better solution.

The result of the FPT experiment can be seen in Fig. 7. We achieved the lowest validation error of 36.15% with $d_{max} = 5$. This was also the best configuration when evaluated using the training sets, which is why this is the configuration that will be used for comparison.

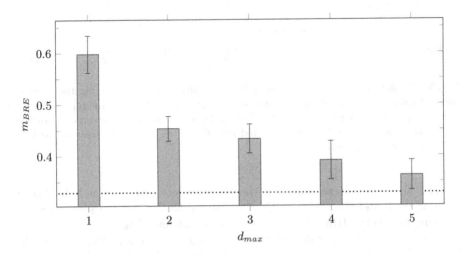

Fig. 7. Evaluation of the FPTs. The horizontal dotted line represents the validation error of the best RF model.

In Fig. 8, a FPT is illustrated that was found for the genre Jazz and the configuration $d_{max} = 3$. Both statements with the features about moods were found to be relevant by the complete search, too.

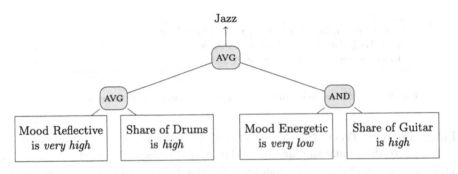

Fig. 8. The FPT built for the genre Jazz and $d_{max} = 3$.

Interestingly, none of the FPTs used any of the negated operators NAND, NOR, or NAVG. This might be explained by the way the FPTs are build. During a leaf extension, it becomes one of the children of the newly inserted fuzzy operator. Inserting a negated operator would reverse the positive impact the leaf holds in the first place. Thus, the algorithm does not select a negated operator.

Evolutionary Approach. For our evolutionary algorithm, there are parameters to control the depth of the randomly created rules (d_{rule}), the number of generations (gen_{max}), the population size ($\#_{pop}$), and the initial number of rules in a rule base ($\#_{rule}$). We used the same $\#_{pop}$ of 100 like in [8]. The parameter d_{rule} was set to 0.25, 0.5, and 0.75, to limit the rule depth to 2, 3, and 4. $\#_{rule}$ was set to 5, 10, and 15. Lastly, gen_{max} needed to be determined. [8] used 80 generations. Since we modified their approach, this would be a good starting point, but we also wanted to try out, whether more or less generations would affect the performance of the model. Therefore, we started the algorithm with $gen_{max} \in \{50, 80, 110\}$. As it turned out, the approach performed better with more generations, and we started the last experiment with 140 generations. The algorithm was executed 15 times for each configuration.

The result of the last experiment can be seen in Fig. 9. Looking at the medians, the lowest validation error of 38.15% was achieved when setting $\#_{rule} = 5$ and $d_{rule} = 0.5$. The overall lowest validation error was 35.59%, using $\#_{rule} = 10$ and $d_{rule} = 0.75$.

When evaluating on the training sets, we obtained the best result with $\#_{rule} = 5$ and $d_{rule} = 0.25$. This is the configuration used for comparison.

When looking at the evolutionary generated rules like in Fig. 10, a few problems become noticeable. Firstly, the algorithm does not pay attention to the complexity of the rules. Rules may be primitive like the first one, or complex like the second one. From generation to generation, they tend to become more complex. Secondly, the structure of the second rule seems to be rather arbitrary. Note, that the fuzzy statement "51 is NOT H" appears three times in the same rule. It is unlikely that this rule needed to be that long. Also, the part "217 is NOT VH NOR 51 is NOT H" can be simplified to "217 is VH AND 51 is H".

Altogether, the algorithm could benefit from mechanisms to control the complexity and construction of the rules. For future work, this can be taken into account by modifying the approach to be multi-objective, where the error and the complexity of the rules should be minimized at the same time.

Comparison of Four Approaches. For each classification method, we determined the configuration with the lowest training error as described in the previous sections. The chosen models were evaluated again on the validation sets, which resulted in the multi-class validation errors summarized in Fig. 11. Overall, we obtained the lowest average validation error with random forest classifier (32.85%) followed by the fuzzy pattern trees (36.15%). Interestingly, the random forest classifier is not the best for every genre. The complete search method had

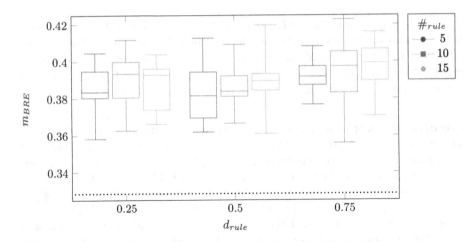

Fig. 9. Evaluation of the evolutionary approach. The whiskers indicate the maximum and minimum, the ends of the box mark the upper and lower quartile, and the middle line stands for the median. The horizontal dotted line shows the validation error of the best RF model.

IF 225 is *VH* THEN Jazz

IF ((51 is NOT *H* NOR (61 is *VL* AVG (51 is NOT *H* NOR (45 is *H* AVG (217 is NOT *VH* NOR 51 is NOT *H*)))))) NOR 54 is NOT *L*) THEN Jazz

---

Fig. 10. Two rules that appeared in the same rule base in one of the experiments. The features referenced by their index to represent the rules shortened. 225: Mood Reflective; 45, 51, 54: Share of Guitar (recognized with different classifiers); 61: Share of Piano; 217: Share of Strings

the highest average error rate (39.77%). Still, it exceeded all other methods for the genre Jazz and the other fuzzy approaches for the genre Rock_and_Pop.

We also compared the algorithm performance by means of the Wilcoxon signed test over categories and folds. The random forest classifier performed significantly better than all fuzzy approaches (p = 0.0020 compared with the complete search, p = 0.0324 with FPTs, and p = 0.0005 with EA). FPTs were significantly better than EA (p = 0.0432). The remaining pairs of algorithms revealed no statistical differences.

Note, that we used the $m = 30$ most relevant rules from the complete search to achieve this result. Since the examined rules were primitive, their premises contained 30 fuzzy statements in total. The best model from the fuzzy pattern trees used the parameter $d_{max} = 5$, which resulted in a complete tree containing only 16 fuzzy statements. Contrary, the number of fuzzy statements used by the evolutionary approach is not controlled in any way. Rules may be primitive or highly complex.

	Best RF	Best Comp	Best FPT	Best Evol
Classical	**.0972**	.1667	.1194	.1233
Electronic_and_Dance	**.1825**	.3263	.2391	.2739
Hip-Hop	.4357	.4596	**.4125**	.4358
Jazz	.4710	**.4652**	.4761	.4907
R_and_B_and_Soul	**.5158**	.6937	.6217	.6749
Rock_and_Pop	**.2685**	.2747	.3000	.3176
Average	**.3285**	.3977	.3615	.3861

Fig. 11. The best validation errors of each method.

5 Conclusion

In this study, we tuned and compared three different approaches for fuzzy rule based classification, one of which uses primitive fuzzy rules while the other two use complex fuzzy rules. Our experiments on the classification into six genres showed that fuzzy rule based approaches are interesting alternatives for classification. However, their interpretability comes at the price of a higher error rate. Fuzzy pattern trees outperformed the other fuzzy methods, with a balanced validation error rate of 36.15% across all genres. Overall, we achieved the best results with the baseline random forest classifier (32.85%). Using fuzzy rule based classifiers in combination with understandable features can make predictions clearer for end users by revealing the rules and indicating their degree of truth for a piece of music. In this scenario, they are more favorable to other classifiers, even if they have a slightly higher validation error.

For future work, in particular the evolutionary approach has room for improvements, as the complexity of the rules was not bound and their construction was uncontrolled. Incorporating these aspects would enforce well-structured rules that may represent the genres better as well. For a fuzzy pattern tree approach, the nondeterministic implementation can be investigated. Further, the definition of linguistic terms can be done in other way than in Fig. 3.

References

1. Berlanga, F.J., del Jesus, M.J., Gacto, M.J., Herrera, F.: A genetic-programming-based approach for the learning of compact fuzzy rule-based classification systems. In: Rutkowski, L., Tadeusiewicz, R., Zadeh, L.A., Żurada, J.M. (eds.) ICAISC 2006. LNCS (LNAI), vol. 4029, pp. 182–191. Springer, Heidelberg (2006). https://doi.org/10.1007/11785231_20
2. Breiman, L.: Random forests. Mach. Learn. J. **45**(1), 5–32 (2001)
3. Fernández, F., Chávez, F.: Fuzzy rule based system ensemble for music genre classification. In: Machado, P., Romero, J., Carballal, A. (eds.) EvoMUSART 2012. LNCS, vol. 7247, pp. 84–95. Springer, Heidelberg (2012). https://doi.org/10.1007/978-3-642-29142-5_8

4. Fernández, F., Chávez, F., Alcalá, R., Herrera, F.: Musical genre classification by means of fuzzy rule-based systems: a preliminary approach. In: Proceedings of the IEEE Congress on Evolutionary Computation, pp. 2571–2577 (2011)
5. Hoffmann, F.: Combining boosting and evolutionary algorithms for learning of fuzzy classification rules. Fuzzy Sets Syst. **141**, 47–58 (2004). https://doi.org/10.1016/S0165-0114(03)00113-1
6. Huang, Z., Gedeon, T.D., Nikravesh, M.: Pattern trees induction: a new machine learning method. IEEE Trans. Fuzzy Syst. **16**(4), 958–970 (2008). https://doi.org/10.1109/TFUZZ.2008.924348
7. Kostek, B., Kaczmarek, A.: Music recommendation based on multidimensional description and similarity measures. Fundam. Inform. **127**(1–4), 325–340 (2013)
8. Linden, R., Bhaya, A.: Evolving fuzzy rules to model gene expression. Biosystems **88**(1), 76–91 (2007). https://doi.org/10.1016/j.biosystems.2006.04.006
9. Salamon, J., Rocha, B., Gómez, E.: Musical genre classification using melody features extracted from polyphonic music signals. In: 2012 IEEE International Conference on Acoustics, Speech and Signal Processing (ICASSP), pp. 81–84 (2012). https://doi.org/10.1109/ICASSP.2012.6287822
10. Senge, R., Hüllermeier, E.: Top-down induction of fuzzy pattern trees. IEEE Trans. Fuzzy Syst. **19**(2), 241–252 (2011). https://doi.org/10.1109/TFUZZ.2010.2093532
11. Senge, R., Hüllermeier, E.: Fast fuzzy pattern tree learning for classification. IEEE Trans. Fuzzy Syst. **23**(6), 2024–2033 (2015). https://doi.org/10.1109/TFUZZ.2015.2396078
12. Vatolkin, I.: Improving supervised music classification by means of multi-objective evolutionary feature selection. Ph.D. thesis, Department of Computer Science, TU Dortmund (2013)
13. Vatolkin, I., Rudolph, G.: Interpretable music categorisation based on fuzzy rules and high-level audio features. In: Lausen, B., Krolak-Schwerdt, S., Böhmer, M. (eds.) Data Science, Learning by Latent Structures, and Knowledge Discovery. SCDAKO, pp. 423–432. Springer, Heidelberg (2015). https://doi.org/10.1007/978-3-662-44983-7_37
14. Yang, Y.H., Liu, C.C., Chen, H.H.: Music emotion classification: a fuzzy approach. In: Proceedings of the 14th ACM International Conference on Multimedia, pp. 81–84 (2006)
15. Zadeh, L.A.: Fuzzy sets. Inf. Control **8**(3), 338–353 (1965). https://doi.org/10.1016/S0019-9958(65)90241-X

Quantum Zentanglement: Combining Picbreeder and Wave Function Collapse to Create Zentangles®

Anna Krolikowski, Sarah Friday, Alice Quintanilla, and Jacob Schrum(✉) ⓘ

Southwestern University, Georgetown, TX 78626, USA
{krolikoa,fridays,quintana,schrum2}@southwestern.edu

Abstract. This paper demonstrates a computational approach to generating art reminiscent of Zentangles by combining Picbreeder with Wave Function Collapse (WFC). Picbreeder interactively evolves images based on user preferences, and selected image tiles are sent to WFC. WFC generates patterns by filling a grid with various rotations of the tile images, placed according to simple constraints. Then other images from Picbreeder act as templates for combining patterns into a final Zentangle image. Although traditional Zentangles are black and white, the system also produces color Zentangles. Automatic evolution experiments using fitness functions instead of user selection were also conducted. Although certain fitness functions occasionally produce degenerate images, many automatically generated Zentangles are aesthetically pleasing and consist of naturalistic patterns. Interactively generated Zentangles are pleasing because they are tailored to the preferences of the user creating them.

Keywords: Zentangle · Compositional Pattern Producing Networks · Wave Function Collapse · Neuroevolution · Interactive evolution

1 Introduction

Artificial intelligence can create unique and visually compelling imagery in a variety of ways. Google Deep Dream [17] and Neural Style Transfer [7] are recent examples demonstrating the power of deep neural networks. However, evolutionary computation has a long history in the field of computer generated art [5,8,10,18]. This paper combines Picbreeder [21], a system based on Compositional Pattern Producing Networks (CPPNs [23]), with Wave Function Collapse (WFC [9]), a procedural content generation method, to create images in the style of Zentangles, a meditative art form (Fig. 1). CPPNs are known for producing naturalistic images. Picbreeder interactively evolves CPPNs taking into account user preferences. Users choose images out of a population to evolve to the next generation. The procedural content generation algorithm WFC is named after a concept from quantum physics, but is a constraint satisfaction algorithm that arranges input tiles into a larger output pattern based on adjacency rules.

© Springer Nature Switzerland AG 2020
J. Romero et al. (Eds.): EvoMUSART 2020, LNCS 12103, pp. 49–65, 2020.
https://doi.org/10.1007/978-3-030-43859-3_4

Fig. 1. Hand-drawn Zentangle. Image has sections with distinct tangle patterns. Art in this style is emulated by the system described in this paper. Credit to Elissa Schrum.

While Picbreeder alone can produce compelling images, it is limited by the activation functions in its CPPNs and the willingness of users to spend time evolving sufficiently intricate networks. WFC adds levels of complexity by introducing small rotational variation and repetition which Picbreeder would have trouble replicating on its own. The resulting output can appear more complex and makes better use of repetition than output produced by Picbreeder alone.

To create Zentangles, images are first evolved in Picbreeder. Images can be black and white (standard for Zentangles) or contain colors. Then the system takes some images, called tile images, and arranges them into patterns using WFC. Specifically, WFC generates rotations and reflections of the tiles, assigns random adjacency rules to them, and uses them to create pattern images. Finally, patterns are combined into a Zentangle according to a partitioning of the space based on one or two template images also taken from the evolved population.

Images for Zentangles can be generated by interactive or automated evolution. Interactively evolved Zentangles reflect the changing experience of a user over time and a user's aesthetic preferences. The user is directly involved in making Zentangles, though randomness influences the outcomes and adds an element of surprise. To automatically evolve Zentangles, the multi-objective evolutionary algorithm NSGA-II [6] was used with a variety of fitness functions. Automatic generation creates a lineage of art from which a user can pick their favorites.

The automatically generated Zentangles prove that human input is not required for the system to produce visually stimulating images. However, the automated system does produce degenerate output on occasion. In contrast, interactively generated Zentangles are based on a user's preferences, making the process more enjoyable to a user and the output possibly more rewarding.

2   Related Work

Computer generated art is a rich and diverse field. Evolution based approaches have a long history in the field [5,8,18]. The approach in this paper is an extension of Picbreeder [21], which uses Compositional Pattern Producing Networks (CPPNs [23]) to generate 2D images. The Picbreeder output is then combined with Wave Function Collapse (WFC [9]) to create Zentangles.

2.1   Art via Compositional Pattern Producing Networks

CPPNs [23] are a generative encoding capable of representing images [21], animations [26], 3D sculptures [4], neural networks [24], and soft-body robots [3]. They are arbitrary topology neural networks evolved via NeuroEvolution of Augmenting Topologies (NEAT [25]) to create images with principles of design seen in nature, such as repetition, repetition with variation, and symmetry. These abilities come from the activation functions CPPNs contain, such as symmetric, asymmetric, and periodic functions. CPPNs work by being repeatedly queried with coordinates in some geometric space. For example, querying x-coordinate values from -1 to 1 and passing them through a Gaussian function results in a symmetric pattern. A full list of available activation functions is in Sect. 3.1.

An early demonstration of CPPNs evolving images was Picbreeder [21], an interactive evolution system. Picbreeder has been extended in many ways. Notably, Endless Forms [4] uses CPPNs to evolve 3D forms made of voxels. Both 2D and 3D animations were generated with AnimationBreeder and 3DAnimationBreeder by adding a time input to CPPNs [26]. A time input has also been used to evolve audio timbres using Breedesizer [13]. CPPNs have also been used in the Procedural Content Generation video games Infinite Art Gallery [12], Artefacts [19], and Petalz [20]. Finally, DrawCompileEvolve uses human-crafted images to initialize a CPPN population, which is then evolved with Picbreeder [28].

2.2   Procedural Content Generation with Wave Function Collapse

Wave Function Collapse (WFC [9]) generates images from a set of tile images by repeating them in a coherent way that takes into account adjacency rules associated with each tile. Named after the phenomenon in quantum physics, WFC is essentially a constraint satisfaction algorithm [15]. Adjacency constraints determine which tiles are allowed to be next to each other, and in which orientation. Tiles are placed probabilistically according to the *minimum entropy* heuristic, which favors placements in locations with fewer available options.

WFC has been used for level generation in the games Proc Skater[1] and Caves of Qud[2]. WFC can also be extended for use on 3D objects and meshes, as well as anything that can be represented by following strict constraints, as shown

---

[1] https://arcadia-clojure.itch.io/proc-skater-2016.
[2] http://www.cavesofqud.com/.

through Martin O'Leary's WFC poetry[3]. O'Leary used syllables as "tiles" and poetic devices as constraints [15]. In creating Zentangles, WFC is a tool to create patterns out of simple tiles that are later placed into a larger composition.

2.3 Zentangle

Zentangle is a meditation-based art form[4]. Creating Zentangles is a mindful, meditative process similar to doodling while practicing Zen meditation [16]. A Zentangle consists of a large shape filled with sections of free-hand patterns called tangles (Fig. 1). Tangles are drawn deliberately and thoughtfully, yet unconcerned with realism or traditional definitions of artistic skill. Zentangle mixes meditation and art in a way that is meant to transcend skill insecurities and foster creativity and introspection. However, because the end results have their own artistic merit, it is interesting to procedurally generate Zentangles, and ignore the meditative aspect of the process.

Computer generated images that look similar include fractal images, reaction-diffusion images [22], and Islamic star patterns [14]. Fractals produce patterns reminiscent of snowflakes, frost, and romanesco broccoli. Bacteria and brain coral growth patterns arise from reaction-diffusion models. Islamic star patterns are rigid and mathematical, an ode to the tile mosaics that the patterns are found in. All of the algorithms produce repetitive yet naturalistic results.

Fractals and Islamic star patterns are intricate, but too repetitive to represent the variation in Zentangles. Reaction-diffusion images have more variety, but still create fairly uniform patterns. The approach to Zentangles in this paper allows for a compelling mix of repetition and variation, both due to the expressive power of CPPNs, and the variety of ways WFC can combine tiles into a pattern.

3 Methods

A Zentangle is made by evolving images using CPPNs, composing them into patterns with WFC, and combining patterns into a final Zentangle. Image evolution can be interactive or automated. All source code is available for download at https://github.com/sarahkfriday/QuantumZentanglement.

3.1 Compositional Pattern Producing Networks

CPPNs have arbitrary topologies and are evolved using NeuroEvolution of Augmenting Topologies (NEAT [25]), which increases the complexity of neural network architectures over generations. The increase in complexity hinges on NEAT's use of structural mutation operators: new neurons can be spliced along existing links, and new links can form between existing neurons. Because CPPNs contain a variety of distinct activation functions, each new neuron has

---

[3] https://libraries.io/github/mewo2/oisin.

[4] https://zentangle.com/.

a randomly selected activation function, although there is also a mutation operation that changes the activation function in an existing neuron. Link weights can also be perturbed via mutation, and the use of historical markings for all new components allows crossover to align components with shared ancestry.

Activation functions within the CPPN's arbitrary topology produce the patterns. The full list of activation functions available are sigmoid, hyperbolic tangent, identity, Gaussian, sine, absolute value, linear piecewise, sawtooth wave, ReLU, softplus, triangle wave, square wave, cosine, and SiL. For many of these functions, multiple versions are available, such as *half* versions whose range is $[0, 1]$ and *full* versions whose range is $[-1, 1]$. Some of these activation functions are simply present because the Picbreeder code used [26] was originally part of a larger, more general neuroevolution system called MM-NEAT[5]. Although use of all of these functions is not necessarily recommended, all are available to humans when interactively evolving images and the specific set of functions used can result in large qualitative differences in the images generated.

CPPNs can function on n-dimensions given n orthogonal inputs [21] as shown through some of Picbreeder's extensions. However, Zentangles are strictly two dimensional. The CPPN is given x and y inputs on a [-1, 1] scale that correspond with every pixel in the image, as well as a d input which represents the distance from the center [21]. Utilizing this tertiary input allows the CPPN to easily create radial patterns. CPPNs encode patterns at infinite resolution, because they can be queried on arbitrarily dense coordinate frames, but the output images in this paper are 1440×1440 pixels. The x, y, and d inputs associated with each pixel are input to the CPPN to determine the color of that pixel.

As in the original Picbreeder, CPPN outputs encode a color using hue, saturation, and brightness. This paper introduces a black and white only option which is different from the grayscale option from the original Picbreeder [21] because *only* black and white are present. These black and white images are produced by using a saturation value of 0, rendering hue irrelevant, and adjusting the brightness values. First, the minimum and maximum CPPN brightness outputs across all pixel coordinates are calculated, in order to calculate the midpoint between them. Next, the actual brightness associated with each pixel is 1 (white) if the original brightness is above the midpoint, and 0 (black) otherwise. The results are dual-toned black and white images, which is the style of traditional Zentangles. In a departure from traditional Zentangles, the system can also create color Zentangle images. To enable switching back and forth between color and black-and-white images during interactive evolution, the hue and saturation outputs of CPPNs are always maintained. However, it is still necessary for template images to have large black regions, so that they can partition the space according to black and non-black regions (see Sect. 3.3). For this reason, only the areas that would be white in a black and white image may contain colors.

---

[5] https://github.com/schrum2/MM-NEAT.

3.2   Wave Function Collapse

Wave Function Collapse is a set of two algorithms, Simple-Tiled and Overlapping, that solve the adjacency constraint problem [9]. Only Simple-Tiled is used in this paper, and it is thus the only algorithm discussed. The algorithm first reads a list of square image tiles and their adjacency constraints. Next, an output array is initialized with each index representing a tile. This array is known as the wave. Elements in the wave are true or false, indicating if the tile is forbidden at that position in the output. The wave is initialized as unobserved, meaning every element is true. Thus begins observation of the wave function. The algorithm selects a tile with the shortest non-empty list of adjacency constraints (lowest nonzero entropy) for that given position in the output. Once selection is final, the tile becomes observed and its information collapsed into the wave by propagation. The algorithm repeats the observation phase until the entire wave is observed returning an output, or a contradiction arises that cannot be resolved.

In Zentangle creation, Picbreeder images are tiles for the Simple-Tiled algorithm. Tiles are rendered by CPPNs at a resolution of 48 × 48 pixels, then randomly assigned symmetry types named after letters associated with symmetry patterns isomorphic to certain characters as shown in Fig. 2. Each symmetry type generates particular rotations and reflections of the tile image, and indicates how those rotations can be placed adjacent to other rotations of the image, or other images. A tile set can consist of rotations and reflections of one or more Picbreeder tiles. Each tile set produces one background pattern via WFC on a 30 × 30 grid, which produces a 1440 × 1440 pixel image. Once two or more pattern images are created from all tile sets, they are combined into a Zentangle.

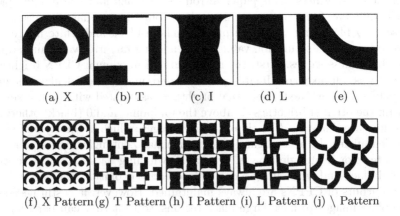

(a) X   (b) T   (c) I   (d) L   (e) \

(f) X Pattern (g) T Pattern (h) I Pattern (i) L Pattern (j) \ Pattern

Fig. 2. Pattern image creation with WFC using black-and-white tiles evolved by Picbreeder. The tile images have been assigned the X, T, I, L and \ symmetry types. Below each tile image is the pattern image created by WFC using the assigned symmetry type. Each type creates specific rotations and reflections. When multiple tiles are present in one pattern, each can have its own symmetry type (not shown).

3.3   Creating Zentangles

Details of Zentangle formation depend on how many images contribute to the Zentangle. The most straightforward example uses three images (Fig. 3). Two are randomly selected as tile images. Each tile is assigned a random symmetry type and creates a pattern via WFC, as described above. The third image is a template, which splits different regions of the Zentangle. The Zentangle is created by analyzing each coordinate location of the template image. If the template is black at that coordinate, it is replaced with the color at that same coordinate from one of the pattern images. If the template is not black at that coordinate, it is replaced with the color at that same coordinate from the other pattern image. The template therefore defines a partitioning of the space in terms of black and non-black with one pattern being mapped to each partition.

Zentangles can also be created with other input image counts (Table 1). In some cases, one image is used as both a tile and a template. If two images are used as templates, then different patterns occupy three regions: black areas in both templates, non-black areas in both templates, and areas black in one but non-black in the other. When six or more images are selected, then individ-

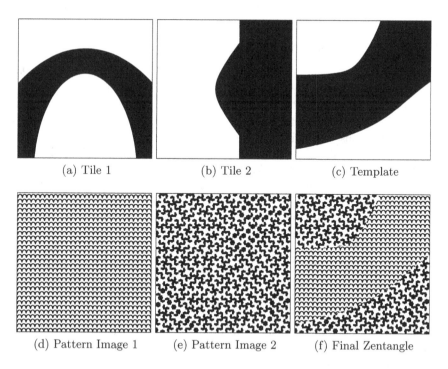

(a) Tile 1	(b) Tile 2	(c) Template
(d) Pattern Image 1	**(e) Pattern Image 2**	**(f) Final Zentangle**

Fig. 3. Zentangle creation from three images. The two images (a) and (b) are randomly assigned to be tile images, so the third image (c) becomes the template. Wave Function Collapse generates pattern image 1 (d) from tile 1 (a) and pattern image 2 (e) from tile 2 (b). The final Zentangle (f) is created by assigning pattern 1 to the black areas of the template image, and pattern 2 to the non-black areas of the template image.

Table 1. Zentangle Construction. When *Total* number of images are selected, some are used as *Templates* and some are used as *Tiles*. In some cases, one image is used as both types. Details in *Additional Information*

Total	Templates	Tiles	Additional Information
2	1	2	One image used as both template and tile
3	1	2	Template and tile images are distinct
4	2	3	One image used as both template and tile. Uses intersection of two templates. Three distinct background patterns
5	2	3	Template and tile images are distinct. Uses the intersection of two templates. Three distinct background patterns
6+	1	5+	Multiple images are used in each of two background patterns

ual background patterns consist of multiple tiles, each with their own random symmetry type. Excess tiles are split evenly across two background patterns.

3.4 Interactive Evolution via Selective Breeding

One way to generate images for Zentangles is interactively, as done in Picbreeder. This selective breeding algorithm uses pure elitist selection. First the user sees $N = 20$ images, then selects $M < N$ individuals as parents for the next generation. These parents are also directly copied to the next generation. Mutation and crossover operations from NEAT are used to create offspring CPPNs. Selection continues in this fashion for as long as the user likes. There is a 50% chance of crossover for each offspring. Whether an offspring has two parents or is a clone, it then has a certain number of mutation chances defined by a user-controlled slider ranging from 1 to 10. For each mutation chance, these rates apply: 30% activation function change rate, 5% per-link weight perturbation rate, 40% link creation rate, and 20% node splice rate. These parameters are the same ones used in the base Picbreeder code [26]. The user also has the option at any point to select images and create a Zentangle out of them, as described above.

3.5 Automated Evolution with NSGA-II

Automated experiments used the multi-objective evolutionary algorithm Non-dominated Sorting Genetic Algorithm II (NSGA-II [6]) to evolve images. NSGA-II uses $(\mu + \lambda)$ selection, specifically $\mu = \lambda = 16$ in this paper, to create new parent populations of size μ from combined parent and child populations of size $\mu + \lambda$. Each child population is created by performing selection on the parent population, and applying crossover and mutation at the same rates used in the interactive experiments (only one mutation chance per offspring). After creating children, the algorithm sorts the combined parent/child population into Pareto layers according to the multi-objective dominance relation, by which one solution is superior to another if it is at least tied in every objective and strictly better in at least one. The first layer is the Pareto front of the population, meaning that it contains no dominated solutions. Removing layers reveals the Pareto fronts of the remaining population members. Elitist selection favors individuals in the less dominated layers, and within layers selection favors solutions that are more distinct from others in their layer in terms of fitness, as determined by a crowding distance metric. Using the crowding distance metric ensures that individuals are evenly spread across the trade-off surface between objectives.

Of the activation functions in Sect. 3.1, the ones used in automated experiments were full sigmoid, full Gaussian, cosine, sine, identity, and half linear piecewise, which were selected because they provide representation of symmetric, periodic, and asymmetric functions. However, other functions could also be used to produce interesting results. For each parent and child population, three Zentangles are created. For each Zentangle, a number between two and six was randomly generated to determine the number of images selected from the population. Those images are then used to compose the Zentangle.

Three different selection schemes were used to evolve images: Random, Half-Black, and Half-Black-3-Colors. The Random scheme selects random images at each generation. The Half-Black scheme favors images whose ratio of black to white pixels is as close to 0.5 as possible. The exact fitness to maximize is $hb(\cdot)$:

$$hb(p) = -\left| \frac{\sum_x \sum_y black(p_{x,y})}{width(p) \times height(p)} - 0.5 \right| \tag{1}$$

where p is a picture, $p_{x,y}$ is the color of a pixel at coordinates (x, y), and $black(p_{x,y})$ is 1 if the designated pixel is black, and 0 otherwise. The Half-Black-3-Colors scheme has four fitness functions: the Half-Black fitness function, and one additional function for each of the RGB color channels. Specifically, the fitness functions for the three channels $red(\cdot)$, $green(\cdot)$, and $blue(\cdot)$ are:

$$red(p) = \frac{sum(p, red) - sum(p, blue) - sum(p, green)}{((width(p) \times height(p)) - \sum_x \sum_y black(p_{x,y})) + 0.0001} \tag{2}$$

$$blue(p) = \frac{sum(p, blue) - sum(p, red) - sum(p, green)}{((width(p) \times height(p)) - \sum_x \sum_y black(p_{x,y})) + 0.0001} \tag{3}$$

$$green(p) = \frac{sum(p, green) - sum(p, blue) - sum(p, red)}{((width(p) \times height(p)) - \sum_x \sum_y black(p_{x,y})) + 0.0001} \tag{4}$$

$$sum(p, c) = \sum_x \sum_y intensity(p_{x,y}, c) \tag{5}$$

where the $sum(p, c)$ for a given color channel c is the sum of the pixel intensity values of that channel across all pixels in p. The $intensity$ function returns the color value in the range $[0, 255]$ of the pixel $p_{x,y}$ for channel c. The value 0.0001 in the denominators is a small term to prevent division by 0. The denominators scale values with respect to the number of non-black pixels to assure that lack of color in the black areas is not punished. Sums from other color channels are subtracted so that each objective is in conflict with the others, encouraging a diversity of colors. These fitness functions do not attempt to evolve to a particular best result, but trajectories where results along the way can be analyzed by a user. Evolutionary runs with each selection scheme were conducted for 50 generations.

4   Results

Both interactively and automatically generated art are discussed individually. All results can be viewed at southwestern.edu/~schrum2/re/zentangle.php.

4.1   Interactively Generated Art

The authors made extensive use of the interactive system. When generating art by selecting images, users maintain influence on the generated Zentangles. Users decide which activation functions are available, and how many mutation chances each offspring has. Although users cannot control the symmetry types assigned to tiles or choose which images become templates, they can control the complexity of Zentangles by deciding which images to include, and how many. Users can also repeatedly generate new Zentangles with the same images, and rely on randomness in the generation process to yield a range of different results.

Users also learn how to generate art that is most appealing to them. By seeing how their selections are used to generate art, they develop an understanding of which image combinations produce what is, in their opinion, the "best" art.

Table 2 shows examples of interactively generated art in black-and-white and color derived from different numbers of images. It demonstrates the variety of results produced by human users. The strict borders between patterns creates a tense interaction between patterns, competing for dominance in pictorial space. The black and white Zentangles contain a depth of space chiseled out by the luminosity of the patterns. Those patterns that are darker recede while those that are lighter advance, dancing through the image. Color Zentangles have a rich balance of complementary and analogous colors which further contrast the individual patterns. Zentangles composed of four to five images entwine patterns in an intricate performance, deepening the pictorial space. Zentangles of six images can be quite noisy, like the black-and-white examples of Table 2. Color examples with six images can also be noisy, but the examples in Table 2 carve out distinct color regions, which gives them distinct forms despite the noise.

Selecting which activation functions can be added to CPPNs results in qualitatively different Zentangles (Table 3). ReLU leads to sharp corners, softplus and Gaussian produce round edges and circles, sawtooth wave creates choppy patterns, and sine with cosine leads to wobbly curves. The user's preferred aesthetic is thus reflected in the end product, making for an enjoyable process.

Table 2. Interactively Generated Zentangle Images. Each row indicates the number of user-selected images that were used to generate the Zentangle. The first two columns show black-and-white images, and the next two columns show color images. This is a small sampling of the range of images humans can generate.

4.2 Automatically Generated Art

While interactive generation of Zentangles yields appealing results, similar results can also be generated automatically. Table 4 contains examples of automatically generated Zentangles similar in quality to interactively generated Zentangles. There is a large range of patterns and colors.

The Random scheme produced these results despite not selecting for any particular features. Images had a wide range of patterns and colors, but occasionally led to homogeneous degenerate Zentangles (Table 5). Zentangles using distinct but similar background tiles can create images where boundaries between patterns are unclear. Because the population is small, distinct offspring may have the same parents and therefore look similar. Still, random selection maintains diversity across generations overall, since it is not honing in on a particular goal.

The Half-Black scheme is meant to assure a good mix of black and non-black regions in each image, since template images depend on this distinction. However, this fitness scheme exerted a high selection pressure, resulting in highly converged populations, leading to more homogeneous degenerates. This kind of convergence is not uncommon in evolutionary art techniques [10,27]. Because there is a specific fitness objective that can be optimized, optimal scoring images quickly take over the population when they emerge. It is apparently easy for CPPNs to generate images with a black bar covering the lower half of the image (Table 5), which is a simplistic pattern. However, WFC's random symmetry assignments can at least make the background patterns interesting in some cases.

The Half-Black-3-Color scheme builds upon the Half-Black scheme by encouraging a variety of color combinations in the white regions of images. This decision maintains more diversity throughout the generations because colors are in competition, and the population is too small to maintain a proper Pareto front across all four objectives. Therefore, this approach keeps producing unexpected designs throughout the entire 50 generation run.

Table 3. Zentangle Images Evolved by Humans with Specific Activation Functions. Each column indicates the only available activation functions used to generate images for the Zentangle, demonstrating qualitative distinctions between activation functions.

ReLU	Softplus & Gaussian	Sawtooth Wave	Sin & Cos

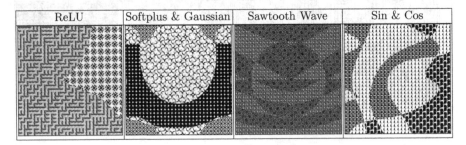

Table 4. Automatically Generated Zentangle Images. Each row indicates the number of images that were used to generate the Zentangle. Each column indicates the fitness scheme which yielded the Zentangle. These automatically generated results are comparable to those generated via interactive evolution.

Table 5. Automatically Generated Degenerate Zentangles. When nearly identical tiles are used to create all background images, the result is a homogeneous pattern, as shown in the first two images. The next two images demonstrate a problem with the Half-Black fitness function, which is easy to optimize: all population members have a black band around one half of the image, though the color portion exhibits variation.

Homogeneous Tiles	Half-Black Template

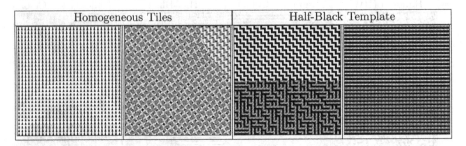

5   Discussion and Future Work

The Zentangle system is able to produce a diverse range of vibrant and interesting images reminiscent of Zentangles. The interactive system shares Picbreeder's ability to adapt to a user's change in aesthetic preferences as the experience progresses. As users evolve images, they may discover their affinity for a curvilinear composition, or a symmetric one. The ability to choose the activation functions for the CPPNs allows for the expression of such preferences. Users also control the complexity level of the resulting Zentangle images via their image choices.

Automated generation is capable of creating similar Zentangles without any human input. However, the automated process occasionally yields degenerate results. Specifically, automated generation can converge to a population of homogeneous images in some cases, as happened with the Half-Black fitness function. Random selection of input images may even pick out similar images from an otherwise diverse population. When multiple background patterns are made from similar tiles, the results are sometimes simplistic.

Some Zentangles composed of six images demonstrate that combining many distinct tiles in one pattern can create a cacophony of line and color. Some of the chaos can be attributed to randomness in assigning symmetry types and image roles. However, appealing compositions sometimes emerge out of the chaos.

Better fitness functions could improve the automatic generation of Zentangles. Fitness functions inspired by Birkhoff's complexity and order measure [2] as well as more recent aesthetic measures could produce a better population of images to create Zentangles from. Basic aesthetic principles worth evaluating include symmetry, repetition, rhythm, and contrast [11], which are elements of design well known within the art community [1]. Heijer and Eiben also proposed several aesthetic measures [10] which could be useful as objectives in multi-objective evolution of images for automated Zentangle creation.

After 50 generations of automatically generating Zentangles, collections of appealing images were obtained with each approach, but results contain several degenerate cases. Half-Black fitness leads to the most homogeneous output, with more diverse images occurring in earlier generations. Random fitness has degenerates sprinkled throughout the generations, but generally maintains high quality and diversity. The Half-Black-3-Color scheme also leads to high-quality results, with images that are even more often filled with diverse colors. Still, it could be beneficial to intelligently select template images from the population. Even in interactive runs specifying the template image would be nice.

Additionally, intelligently configuring WFC may create more structured patterns. For example, some tiles may lend themselves to certain symmetry types. The ability to intelligently associate tiles with symmetry types could create patterns with improved balance, rhythm, and repetition. Alternatively, the user could select the symmetry type for each tile. Of course, demanding too much input from the user could make the system tedious to use and prevent the serendipitous discovery of unexpected output. Similarly, although it may be possible to define criteria for which images to evolve and how to combine them into Zentangles, doing so successfully may take more effort than it is worth since the results become more restricted as the specificity of the fitness functions grows. The results from the Random fitness function are already interesting and pleasing in most cases, so increasing the expressive power of the system may be more valuable than strictly controlling the selection process.

There remain a number of differences between hand-drawn Zentangles and the Zentangles produced in this paper. One difference is variation in pattern scale that can be found in hand-drawn Zentangles. The current system lacks scale variation, but this feature should be easy to add because CPPNs can render images at arbitrary resolutions [23]. Hand-drawn Zentangles can also have patterns in arbitrary orientations, which is another enhancement that should be easy to add. Allowing for more than three background patterns is yet another feature that would make generated results similar to hand-drawn ones. Perhaps the most challenging types of patterns from hand-drawn Zentangles for this system to produce are organic patterns not produced by repetition of tile-like elements. However, a hybrid system that incorporates art from other systems, such as the reaction-diffusion images mentioned earlier [22], could provide the occasional organic pattern. Using WFC's Overlapping algorithm instead of the Simple-Tiled algorithm also has potential to make more organic patterns [9].

Whether the generated Zentangles are less visually appealing than hand-drawn Zentangles is a matter of subjectivity. However, the proposed enhancements should at least produce results closer to what humans can draw.

6   Conclusion

Through the combined use of Picbreeder and Wave Function Collapse, a computer can generate complex, Zentangle-esque art. In fact, a human user is not even needed to produce the images, though human input can be valuable, since humans recognize which images might combine to produce the best results. Automatic generation of Zentangles utilizing three different fitness functions has proven its ability to produce intriguing results as well, despite some degenerate output. However, further development should strive to close the qualitative gap between evolved and hand-drawn Zentangles, by allowing for more diverse pattern types and ways of combining them. Such progress may arise from combining multiple techniques into hybrid systems, as was done in this paper.

References

1. Arnheim, R.: Art and Visual Perception: A Psychology of the Creative Eye. University of California Press, Berkeley (1954)
2. Birkhoff, G.: Aesthetic Measure. Harvard University Press, Cambridge (1933)
3. Cheney, N., MacCurdy, R., Clune, J., Lipson, H.: Unshackling evolution: evolving soft robots with multiple materials and a powerful generative encoding. In: Genetic and Evolutionary Computation Conference. ACM (2013)
4. Clune, J., Lipson, H.: Evolving three-dimensional objects with a generative encoding inspired by developmental biology. In: European Conference on Artificial Life, pp. 141–148 (2011)
5. Dawkins, R.: The Blind Watchmaker. Longman Scientific and Technical, Harlow (1986)
6. Deb, K., Pratap, A., Agarwal, S., Meyarivan, T.: A fast and elitist multiobjective genetic algorithm: NSGA-II. IEEE Trans. Evol. Comput. **6**, 182–197 (2002)
7. Gatys, L., Ecker, A., Bethge, M.: A neural algorithm of artistic style. J. Vis. **16**(12), 326 (2016)
8. Greenfield, G.: Robot paintings evolved using simulated robots. In: Rothlauf, F., et al. (eds.) EvoWorkshops 2006. LNCS, vol. 3907, pp. 611–621. Springer, Heidelberg (2006). https://doi.org/10.1007/11732242_58
9. Gumin, M.: WaveFunctionCollapse. GitHub repository (2016). https://github.com/mxgmn/WaveFunctionCollapse
10. den Heijer, E., Eiben, A.E.: Comparing aesthetic measures for evolutionary art. In: Di Chio, C., et al. (eds.) EvoApplications 2010. LNCS, vol. 6025, pp. 311–320. Springer, Heidelberg (2010). https://doi.org/10.1007/978-3-642-12242-2_32
11. Hoenig, F.: Defining Computational Aesthetics. Computational Aesthetics in Graphics, Visualization and Imaging, pp. 13–18 (2005)
12. Hollingsworth, B., Schrum, J.: Infinite art gallery: a game world of interactively evolved artwork. In: IEEE Congress on Evolutionary Computation (2019)
13. Jónsson, B.T., Hoover, A.K., Risi, S.: Interactively evolving compositional sound synthesis networks. In: Genetic and Evolutionary Computation Conference, pp. 321–328. ACM (2015)
14. Kaplan, C.S.: Computer generated Islamic star patterns. In: Bridges 2000: Mathematical Connections in Art, Music and Science, pp. 105–112 (2000)
15. Karth, I., Smith, A.M.: WaveFunctionCollapse is constraint solving in the wild. In: Foundations of Digital Games. ACM (2017)
16. Kopeschny, D.A.: The phenomenological experience of Zentangle and the implications for art therapy. Master's thesis, St. Stephen's College, Alberta, Canada (2016)
17. Mordvintsev, A., Olah, C., Tyka, M.: Inceptionism: Going Deeper Into Neural Networks (2015). https://research.googleblog.com/2015/06/inceptionism-going-deeper-into-neural.html

18. Neumann, A., Szpak, Z.L., Chojnacki, W., Neumann, F.: Evolutionary image composition using feature covariance matrices. In: Genetic and Evolutionary Computation Conference. ACM (2017)

19. Patrascu, C., Risi, S.: Artefacts: minecraft meets collaborative interactive evolution. In: Computational Intelligence and Games, pp. 349–356. IEEE (2016)

20. Risi, S., Lehman, J., D'Ambrosio, D.B., Hall, R., Stanley, K.O.: Petalz: search-based procedural content generation for the casual gamer. IEEE Trans. Comput. Intell. AI Games **8**, 244–255 (2015)

21. Secretan, J., et al.: Picbreeder: a case study in collaborative evolutionary exploration of design space. Evol. Comput. **19**(3), 373–403 (2011)

22. Sims, K.: Interactive evolution of dynamical systems. In: European Conference on Artificial Life, pp. 171–178 (1992)

23. Stanley, K.O.: Compositional pattern producing networks: a novel abstraction of development. Genet. Program. Evolvable Mach. **8**(2), 131–162 (2007). https://doi.org/10.1007/s10710-007-9028-8

24. Stanley, K.O., D'Ambrosio, D.B., Gauci, J.: A hypercube-based encoding for evolving large-scale neural networks. Artif. Life **15**, 185–212 (2009)

25. Stanley, K.O., Miikkulainen, R.: Evolving neural networks through augmenting topologies. Evol. Comput. **10**, 99–127 (2002)

26. Tweraser, I., Gillespie, L.E., Schrum, J.: Querying across time to interactively evolve animations. In: Genetic and Evolutionary Computation Conference (2018)

27. Woolley, B.G., Stanley, K.O.: On the deleterious effects of a priori objectives on evolution and representation. In: Genetic and Evolutionary Computation Conference, pp. 957–964. ACM (2011)

28. Zhang, J., Taarnby, R., Liapis, A., Risi, S.: DrawCompileEvolve: sparking interactive evolutionary art with human creations. In: Johnson, C., Carballal, A., Correia, J. (eds.) EvoMUSART 2015. LNCS, vol. 9027, pp. 261–273. Springer, Cham (2015). https://doi.org/10.1007/978-3-319-16498-4_23

Emerging Technology System Evolution

Matthew Lewis[(⊠)] [iD]

Ohio State University, 1813 N High Street, Columbus, OH 43210, USA
Lewis.239@osu.edu

Abstract. This paper introduces a strategy for exploring possibility spaces for creative design of emerging technology systems. These systems are represented as directed acyclic graphs with nodes representing different technologies and data or media types. The naming of system and data components is intended to be populated by a subjective, personal ontology. The source material can be simply modified by a user who wishes to explore a space of possibilities, constrained by specific resources, to generate simple visual descriptions of systems that can be quickly browsed. These dataflow visualizations are a familiar representation to those who use visual programming interfaces for creative media processing. An interactive genetic programming based approach is used with small populations of individual designs. The system is being developed using web browser technology. A minimum functionality proof of concept system has been implemented. Next steps and current challenges are discussed.

Keywords: Emerging technologies · Creative systems · Interactive evolution

1 Introduction

It is extremely common for designers working with emerging technologies to find themselves brainstorming ideas within possibility spaces of systems designs. Such explorations may be driven by the availability of specific resources, funding opportunities, or learning goals. It is often stated that good design should focus on creating solutions to problems and specific needs. However, surprisingly frequently, individuals find themselves searching for applications, given a set of capabilities and resources, i.e., "solutions in search of a problem."

During such ideation sessions, alone or with a group, there is often a great deal of discipline-specific and contextual shorthand used. Important terminology is connected by arrows as recorded fragments, frequently on a whiteboard. These resulting diagrams generated by the group or individual refer to larger ideas, quickly summarized for concise visual consideration.

Individuals working in interdisciplinary creative environments who participate in such efforts acquire significant expertise with the capabilities and requirements for integrating different technologies. When collaborators, students, or colleagues wrestle with solutions and constraints, the pros and cons of many "what if...?" ideas are rapidly considered and discarded. A few examples from recent collaborative brainstorming sessions involving emerging technologies have included: smart mobility and

© Springer Nature Switzerland AG 2020
J. Romero et al. (Eds.): EvoMUSART 2020, LNCS 12103, pp. 66–79, 2020.
https://doi.org/10.1007/978-3-030-43859-3_5

health apps, drone safety and 3D scanning, augmented reality and environmental shelter prototyping, and IoT interfaces and creative performance.

Partial ideas are frequently discussed via short descriptions based on system components and capabilities: *"What if we used a set of drones to stream video from different perspectives?" "Could we reconstruct the video streams into 3D models?" "What if we filtered the model and projected it back onto the surface with projection mapping?"* When ideas within a space of possibilities flow freely, innovation can be considered in terms of what's referred to as the "adjacent possible" [1]. Designers working creatively navigate through such adjacent spaces of partial ideas to explore system design possibilities as they sketch and prototype, gradually arriving at solutions.

Generative design approaches have been implemented in fields as diverse as music, painting, architecture, and dance. Instead of crafting each possible solution "by hand" for consideration, computer software has been developed to produce possibilities for a designer whose role becomes more akin to a curator. The work being developed that is described in this paper considers the possibility of applying such techniques to the creative generation of technological system descriptions.

Because an interactive creative evolutionary design approach is being employed, there are three primary goals: First, the population of proposed designs should be extremely simple to quickly browse. Second, the representation of the elements of the possibility space need to be trivial to customize based on an individual user's subjective experience, knowledge, and interests. Finally, the implementation approach should be as simple and practical as possible.

The usage of such systems should help designers think about education with emerging technologies, just as those in the field of design education have stated to be so critical [2]. At the same time, we must always be considering and evaluating potential negative impacts of emerging technologies [3].

1.1   Overview

This paper is structured to present current progress applying a standard interactive evolutionary design approach to a specific creative design problem domain. After a brief discussion of prior work, the representation selected for the domain will be discussed. The evolutionary approach selected (based on interactive genetic programming systems) will be considered. Initial random individuals and populations of varying complexity will be shown to get a quick sense of the solution space. The domain-specific details of the mating and mutation of individuals will be discussed. Finally, the success of evolved populations will be considered.

1.2   Prior Work

The earliest interactive evolutionary design systems also evolved populations of branching structures [4, 5]. However, as aesthetic drawings or forms, their intent was very different than the simple directed dataflow graphs presented here. While those early works relied on genetic algorithms, genetic programming (GP) approaches have also been used in early interactive evolution systems ever since Sims used them to generate images [6]. GP form generation systems were also used in early interactive VR environments [7].

Roughly at the same time as the beginning of interactive evolutionary design, early AI expert systems were being used to generate visual representations of technology system components and their relationships for training and related uses [8]. There are now decades of research on formal knowledge representation infrastructure (e.g., RDF and OWL) useful for building scalable and robust ontologies of systems for applications like the semantic web [9]. The power and complexity of such efforts are far beyond the modest knowledge representation needs of the work described in this paper. The design goals for this work are additionally biased towards simple, easy to understand systems in order to improve general accessibility for target users such as non-programmer design students beginning to work with emerging technologies.

There have also been approaches to representing technological systems in other creative contexts. In the art and technology community, Jim Campbell's well known "Formula for Computer Art" animated diagram and associated paper concisely visualize the idea of inputs and outputs between different types of technologies [10, 11]. In the design research community, Tow has described conceptually positioning technologies in a cartesian space, with axes derived from properties of the technologies to be represented, to recognize "voids" (and thus design opportunities) [12]. Using the properties of technologies in this way (i.e., discovering unknown structures by implicitly defining their relationships) blurs Carvalhais' binary distinction between descriptive and generative artistic processes [13].

Finally, physical cards have long been used by designers for ideation in grid or domino-style layouts representing physical computing and interactive systems [14–17]. There have been several projects that correspondingly support actual physical prototyping of functional systems, using modular blocks with electronic components. These components can be connected in order to prototype different sequences of inputs and outputs [18, 19]. The work described in this paper is intended as a virtual middle ground between such paper and electronic modular system component representations.

2   Representation

Technology-based systems are commonly described in terms of their components and the relationships between components: "a robot uses a camera to find people and turns towards them to answer their questions using a database of facts." While sentences of just a few lines can often concisely represent such systems, another representation lends itself more easily to rapid comparison of multiple systems. Simple system diagrams, represented by basic shapes, annotations, and connecting arrows, can be compared side-by-side to easily spot similarities and differences.

The components in the systems below have been visually represented as either a "technology" or a "datatype" although both of these terms are being used extremely imprecisely. The "technologies" may be anything that can be considered as accepting inputs and producing outputs. The "datatypes" are any material (information, media, etc.) being passed between technologies. As stated above, while there are myriad approaches to representing conceptual relationships such as these unambiguously, the ambiguity here is intended as a usability feature. A user should not have to think too much about the descriptive keywords chosen, and it shouldn't matter if the creative system generates occasional nonsensical system relationships as a result.

As opposed to requiring the navigation of a database interface (and supporting the related infrastructure) a simple text file that lists collections of keywords can be modified in any text editor. The YAML standard is used for easy human readability and more forgiving editing due to minimal punctuation requirements than alternatives such as JSON (to which it can be easily converted). Below is a small subset of roughly fifty technologies thus represented and used to generate the system diagrams shown in the remainder of this paper:

```
camera:
   i: [light, print, object, animation, video, visual,
      image, text, direction, sculpture]
   o: [image, video, color, location]
drone:
   i: [control, direction, location]
   o: [image, video, motion]
HMD:
   i: [video, image, geometry, VE, animation]
   o: [light, video, image]
IoT:
   i: [motion, presence, light, sound, temperature]
   o: [data, number]
projector:
   i: [image, video, map, animation]
   o: [projection, image, light, video, color]
voice_assistant:
   i: [voice, task, identity, decision]
   o: [control, music, sound, voice, fact]
```

After providing the name for a technology, sets of input and output keywords are provided within brackets. When undesirable unintentional connections between inputs and outputs inevitably emerge, the simplicity of the grammar makes it easy to delete, amend, or adjust these technology descriptions. This makes it very easy to experiment, personalize, and shape the search space quickly.

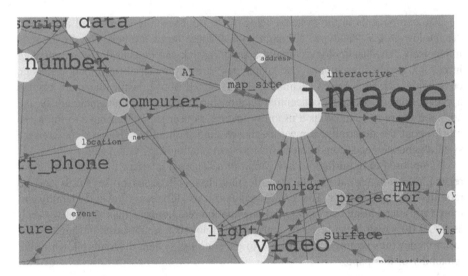

Fig. 1. Detail of a dense graph representation of the connected technologies and datatypes.

By using the same terms in the input and output sets of different technologies, a network of connected technologies is implicitly created. A detail of a dense visualization of this network is shown in Fig. 1. Representing only tens of technologies and datatypes with hundreds of connections between them generates a space with thousands of small subgraphs to be considered. The system diagrams shown in the remainder of this paper are subgraphs from this space. Instead of attempting to interactively evaluate these technology system subgraphs within the context of the arbitrarily complex network of Fig. 1, we can view them instead as individual dataflow system diagrams, as shown in Fig. 2.

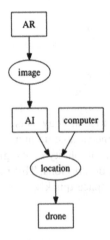

Fig. 2. Example system diagram.

The system example shown in Fig. 2 could also be represented by a text description such as, "Augmented reality software produces images that are interpreted by AI software. The AI (and another computer) then generate locations that are used to navigate a drone." As with generative systems that produce images or sound, the *interpretation* of the generated media depends on the experiences and knowledge of the viewer. Also, as with other products of interactive evolutionary design systems, the complexity of the generated artifacts dictates how quickly they can be subjectively understood and evaluated. Music evolution systems provide a classic example. It is very challenging to listen to several generated pieces, while holding each in mind to compare to the subsequent instances.

Figure 3 shows a minimal system on the left, which can be understood at a glance as "a 3D printer generates an object that is viewed by a camera". Many such simple individual system descriptions can be evaluated quickly when displayed as a population of systems. The randomly generated system on the right however is significantly more complex, taking a much longer time to interpret. It could be interpreted as a voice recognition system that stores text in a blockchain, while also using it to control a drone. Additionally, a sensor of some sort and an "Internet of Things" device feed locations to a drone. The IoT device also reports identity information to somebody. As systems grow more complex, they also become more difficult to hold in mind in their entirety or to generate narratives about their possible usage and impact. The system for visually monitoring 3d printer progress on the left, is much easier to conceptualize and describe than the system on the right ("Verifiable voice-controlled identity-based sensor network drones").

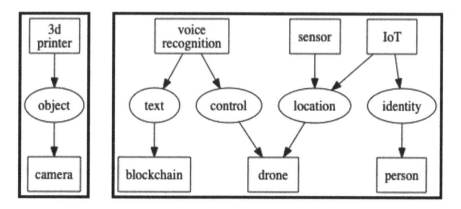

Fig. 3. Minimal system diagram (left), complex system diagram (right).

3   Interactive Evolution

3.1   Initial Populations

Given the above representation, it is straightforward to generate random individual system descriptions. A single linear random *link* is first created between two technologies: a technology T1 is randomly selected, one of its possible datatype outputs D1 is randomly chosen, and finally a second technology T2 is then chosen which has output D1 in its list of possible inputs: T1 → D1 → T2.

The system can then *grow* this simple link by choosing one of its three nodes at random, then sprouting a new link either "forwards" or "backwards" from the selected node in the directed graph. For example, if the technology node T1 was randomly selected, this means either outputting a second datatype D2, or else creating a new input link D0 "behind" it, i.e., T0 → D0 → T1 → D1 → T2. Note that growing a new input or output datatype node from a technology node also requires growing an additional technology node from the new datatype, e.g., T0 → D0 in the previous example. Nodes can be grown a configurable number of times which determines the complexity of the initial population. In Fig. 3, the minimal system on the left shows a basic link between two technologies with no additional node growth. The system on the right was grown four times.

The basic techniques frequently used for genetic programming were chosen as an appropriate fit given the simple branching hierarchical structure of the systems being represented. The individuals being evolved are not strictly *programs*, so "genetic programming" might not be a technically accurate description. However, as the systems represent dataflow processing networks, corresponding GP-based approaches for graph mutation and mating have been implemented and will be described below, with some domain specific details mentioned.

3.2   Mutation

Mutation is applied to a system one of three ways. First, a leaf technology node (i.e. one with either no inputs or no outputs) can be *removed*. Its connected datatype node will also be removed if it was only connected to the removed technology. Second, any node can be *grown* by adding randomly connected new nodes using the strategy described in the previous Sect. 3.1.

The third method involves *replacing* one of the nodes with a new node that can fit into its place. One node can be substituted for another if *both* inputs and outputs match. This is simpler for a leaf node that is connected to just one datatype (by input or output). There are likely several other technologies that also could connect to that same datatype. It may be impossible however for an *interior* node with several input and output connections to find a similarly connectable technology or datatype. Figure 4 shows possible valid replacement mutations generated for both leaf and interior nodes for the system shown in Fig. 2.

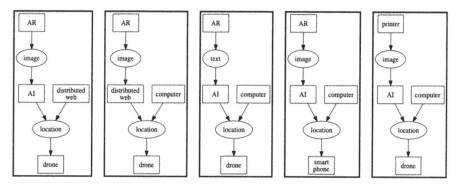

Fig. 4. Possible replacement mutations of leaf or interior system nodes.

3.3 Mating

While more complex strategies are possible, only a basic graph mating approach has been implemented. Given a pair of systems, a random node N is selected in the first system. Every node in the second system is considered to see whether it could possibly be connected below node N (with matching inputs and outputs.) If N is a technology, then only datatype nodes in the second system that are in N's list of possible outputs are feasible. If N is a datatype then only technology nodes in the second system that have datatype N as a possible input type could connect. After the list of all possible connection splicing points is created, one is chosen at random. Figure 5 shows seven possible offspring systems generated at different splicing points from the two parents on the top left.

3.4 Evolving Populations

The evolution interface is at a functional proof-of-concept stage. For implementation prototyping, the above algorithms were developed first as command line Node.js scripts, and then as a local Node.js server that communicates with a local web browser. A randomly generated initial population of individual systems is first displayed in the web browser when it connects to the server. The most subjectively interesting system designs can be selected by clicking on them. Selection feedback is provided through border highlighting. Clicking on a previously selected system deselects it. A button on the web page can be pressed when done to finalize the mating pool selection and generate a new population of offspring.

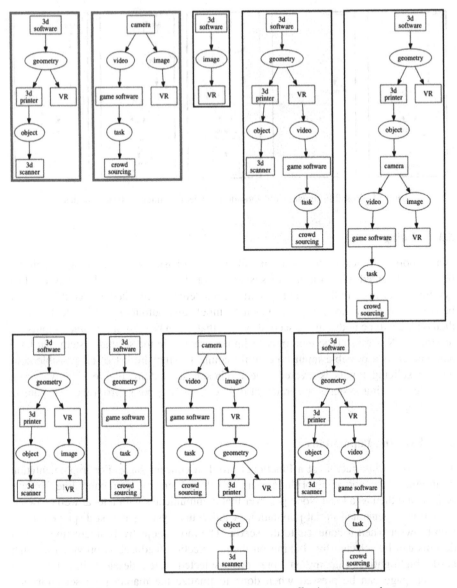

Fig. 5. Mating the top left pair of systems; seven offspring systems shown.

Figure 6 shows a random initial population of fifteen individual system designs with each system only grown at most by one or two additional links. Figure 7 shows a population four generations later. Offspring are generated by mating random pairs from the user selected mating pool followed by a single mutation applied to each offspring. When only a single system is selected, systems are only mutated. Duplicate systems are identified when generated and replaced with a unique offspring when possible.

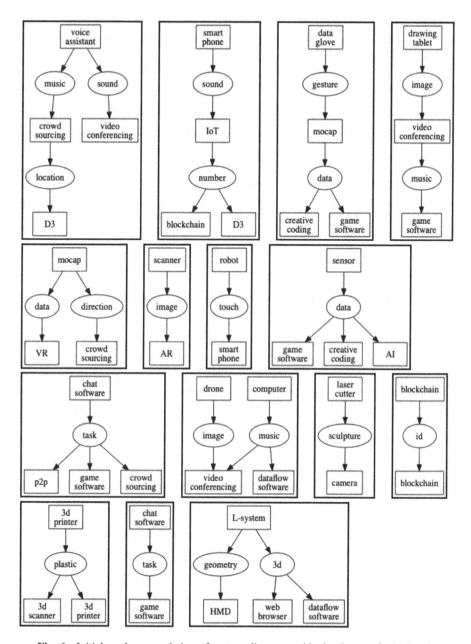

Fig. 6. Initial random population of system diagrams with simple complexity level.

A necessary design goal for any evolutionary design system is for offspring to have a reasonable possibility of inheriting the attributes that were responsible for the selected parents' higher fitness. If there is no inheritance of phenotypic fitness qualities, then offspring will be no better than randomly generated individuals. If phenotypic fitness qualities *can* be inherited, then it follows that the offspring that do inherit such qualities should have higher fitness than randomly generated individuals with average fitness.

Phenotypic attributes that exhibit higher fitness in *this* domain are typically novel and/or interesting combinations of connected technologies. The mating procedure used frequently generates offspring that inherit the unique combination that caused the individual's selection. This proof-of-concept system therefore can consistently meet the minimal requirement of better-than-random results.

A significant further consideration when evaluating the generative systems is inevitably the richness of their possibility space. This may be partially based on training data, generative functions, representations, etc. But with human-in-the-loop systems, it is also dependent on the interpretations of the people using the system, as well as the match between the capabilities of the system and the user. A creative system developed according to the needs, experiences, and preferences of one artist or designer may not be appropriate for those of another.

The current output generated by this system meets the *technical* goals described above: A collection of technology components and their capabilities has been represented in a way that is simple to create and modify. The collection can be used to generate and display a population of random novel system descriptions. These systems can be quickly evaluated and selected using a traditional interactive evolutionary design interface.

When evaluating generative systems, there is occasionally a mistaken expectation that the software was developed as a tool for searching for a solution to a specific problem. For example, "Is this new tool faster than an alternative approach for finding the best solution?" In many disciplines such as the arts, business, or humanities, scenarios requiring creativity include problem discovery, or matching capabilities to problems. In any discipline that values idea generation, tools that help to do so have the potential to be very useful. Creative systems can assist with generating ideas, not just solving problems.

Initial use of the system described in this paper has revealed significant room for improvement of offspring which should be attainable through the following modifications. While basic graph splicing approaches used for genetic programming can be relatively simple to implement, high fitness qualities are often not inherited by offspring due to indiscriminate splicing of parents, e.g., when the top 10% of one parent is spliced with the bottom 10% of another parent, removing the majority of both parents' phenotypic attributes from the offspring. Mating two graphs can be done more intelligently by biasing splice point selection towards the center of the parents. This can also better maintain the overall size of individuals (e.g., if using 75% of one parent, try to use roughly 25% of the other.) Furthermore, indiscriminate splice point selection often causes too much system growth. Splice selection could be biased towards reducing system size to keep diagrams able to be rapidly browsed (even allowing the option of user configurable graph complexity biasing).

Finally, balancing the perceived impact of mutation and mating seems like a strategy worth investigating in the future. This should give greater control over the speed of movement through the possibility space.

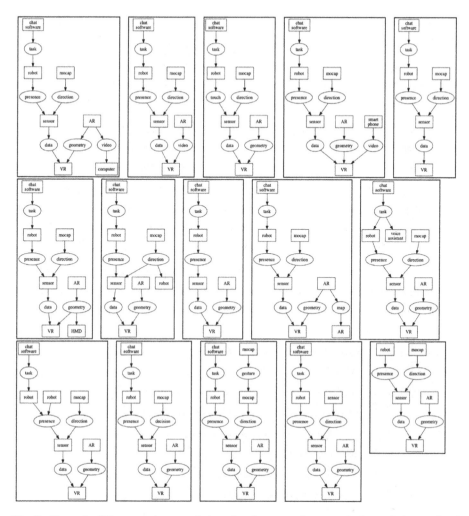

Fig. 7. Example fifth generation population showing emerging complexity and the result of mutating a single selected system.

4 Next Steps

Exploration of this possibility space, its limitations, and its inherent properties is only just beginning. After the mating and mutation algorithms have been adjusted to beneficially bias the size and attribute inheritance of offspring, as discussed above, user testing will be conducted within the local designer community. While this work is motivated by the author's professional creative design needs and research interests, the system architecture decisions are being made with the intent to be accessible to others who might also find this strategy useful. The ease (or difficulty) designers have in assembling their own subjective set of building blocks for their individual problem domains will be of particular interest.

Finally, there are three additional areas for future work: The first is the use of arbitrary text metadata, beyond just inputs and outputs. Qualitative subjective terms related to cost, sustainability, accessibility, privacy, and so forth can be associated with the different technologies and datatypes (e.g., tagging cameras, drones, and computer vision with words like "privacy" or "surveillance"). This would let users explicitly represent additional information that they could then use to limit and/or bias system generation and selection processes.

Second, because of the underlying text-based representation, it should be possible to generate audible text descriptions for improved accessibility of the interface, applying universal design principles to interactive evolutionary design. Because the system is being developed for web browsers using open standards, integration with existing screen reader software should be feasible.

Finally, a longer-term goal is to visually represent generated systems with virtual prototypes. Given visual and interactive analogues of technologies such as 2D icons and 3D models, visual representations of the systems described could be generated within interactive virtual spaces (Fig. 8). This project is one component of a system being developed to allow prototyping and evaluation of responsive environment concepts within immersive virtual environments. Interactive evolution has been proposed as one method of exploring possibility spaces within specific spatial contexts.

Fig. 8. VR environment of a physical lab with virtual projectors and 360 images.

The core ideas behind this project were outlined and presented in simpler form in a position paper over a decade ago [20]. They are only being implemented now that VR technology is serving as a motivator, allowing virtual simulation of interactive technology systems. It is questionable whether current emerging technologies are any easier to work with than those of the past. Technological acceleration appears very real however and, as such, the need for tools that can assist with creative design efforts seems more important than ever.

References

1. The Adjacent Possible. https://www.edge.org/conversation/stuart_a_kauffman-the-adjacent-possible. Accessed 13 July 2019
2. Armstrong, H., et al.: AIGA Designer 2025: Why Design Education Should Pay Attention to Trends (2017). https://educators.aiga.org/wp-content/uploads/2017/08/DESIGNER-2025-SUMMARY.pdf
3. Zuboff, S.: The Age of Surveillance Capitalism: The Fight for a Human Future at the New Frontier of Power. PublicAffairs, New York (2019)
4. Dawkins, R.: The Blind Watchmaker: Why the Evidence of Evolution Reveals a Universe Without Design. Norton, New York (1986)
5. Todd, S., Latham, W.: Evolutionary Art and Computers. Academic Press, London; San Diego (1992)
6. Sims, K.: Artificial evolution for computer graphics. In: Proceedings of the 18th Annual Conference on Computer Graphics and Interactive Techniques, pp. 319–328. ACM, New York (1991). https://doi.org/10.1145/122718.122752
7. Das, S., Franguiadakis, T., Papka, M., DeFanti, T.A., Sandin, D.J.: A genetic programming application in virtual reality. In: Proceedings of 1994 IEEE 3rd International Fuzzy Systems Conference, vol. 3, pp. 1985–1989 (1994). https://doi.org/10.1109/FUZZY.1994.343536
8. Feiner, S.K., McKeown, K.: COMET: generating coordinated multimedia explanations. In: CHI (1991). https://doi.org/10.1145/108844.108997
9. Berners-Lee, T., Hendler, J., Lassila, O.: The semantic web in Scientific American. Sci. Am. Mag. **284**, 34–43 (2001)
10. Campbell, J.: Portfolio: Miscellaneous References. http://www.jimcampbell.tv/portfolio/miscellaneous_references/. Accessed 13 Nov 2019
11. Campbell, J.: Delusions of dialogue: control and choice in interactive art. Leonardo **33**, 133–136 (2000). https://doi.org/10.1162/002409400552397
12. Tow, R.: Strategy, tactics, and heuristics for research: a structuralist approach. In: Laurel, B. (ed.) Design Research: Methods and Perspectives. MIT Press, Cambridge (2003)
13. Carvalhais, M.: Artificial Aesthetics: Creative Practices in Computational Art and Design. U. Porto Edições, Porto (2016)
14. Turkel, W.J.: Digital History Hacks (2005–2008): Physical Computing Cards. http://digitalhistoryhacks.blogspot.com/2007/11/physical-computing-cards.html. Accessed 12 Nov 2019
15. Saffer, D.: Designing Gestural Interfaces: Touchscreens and Interactive Devices. O'Reilly Media Inc., Newton (2008)
16. Golembewski, M., Selby, M.: Ideation decks: a card-based design ideation tool. In: Proceedings of the 8th ACM Conference on Designing Interactive Systems, pp. 89–92. ACM, New York (2010). https://doi.org/10.1145/1858171.1858189
17. Compton, K., Melcer, E., Mateas, M.: Generominos: ideation cards for interactive generativity. In: Thirteenth Artificial Intelligence and Interactive Digital Entertainment Conference (2017)
18. Interstacks—MAYA Design. http://maya.com/work/interstacks. Accessed 11 Nov 2019
19. littleBits—Electronic Building Blocks for the 21st Century. https://littlebits.com/. Accessed 12 Nov 2019
20. Lewis, M.R.: Casually evolving creative technology systems. In: Boden, M., D'Inverno, M., McCormack, J. (eds.) Computational Creativity: An Interdisciplinary Approach. Schloss Dagstuhl - Leibniz-Zentrum fuer Informatik, Germany, Dagstuhl (2009)

Fusion of Hilbert-Huang Transform and Deep Convolutional Neural Network for Predominant Musical Instruments Recognition

Xiaoquan Li, Kaiqi Wang, John Soraghan, and Jinchang Ren[✉]

Department of Electronic and Electrical Engineering, University of Strathclyde,
Royal College Building, 204 George Street, Glasgow G1 1XW, UK
jinchang.ren@strath.ac.uk

Abstract. As a subset of music information retrieval (MIR), predominant musical instruments recognition (PMIR) has attracted substantial interest in recent years due to its uniqueness and high commercial value in key areas of music analysis such as music retrieval and automatic music transcription. With the attention paid to deep learning and artificial intelligence, they have been more and more widely applied in the field of MIR, thus making breakthroughs in some sub-fields that have been stuck in the bottleneck. In this paper, the Hilbert-Huang Transform (HHT) is employed to map one-dimensional audio data into two-dimensional matrix format, followed by a deep convolutional neural network developed to learn affluent and effective features for PMIR. In total 6705 audio pieces including 11 musical instruments are used to validate the efficacy of our proposed approach. The results are compared to four benchmarking methods and show significant improvements in terms of precision, recall and F1 measures.

Keywords: Predominant musical instrument recognition (PMIR) ·
Convolutional neural network (CNN) · Hilbert-Huang Transform (HHT)

1 Introduction

Music information retrieval (MIR) has drawn significant research attention in the last decade and has been successfully applied in many applications, such as music retrieval and automatic music transcription [3]. Instead of manually identifying the rhythm, genre and timbre by our ears, MIR techniques can automatically label audio data based on their time and frequency information. As a sub-task of MIR, predominant musical instrument recognition (PMIR) enables customers to search music by instruments, as well as making music transcription easier and more accurate [6]. However, PMIR is a very challenging topic and current PMIR approaches have yet to be commercialised due mainly to the lack of robust performance. However, it is quite useful in some applications such as assisting automatic music transcription (AMT) detection, crude instrument classification, and instrument characterisation.

The identification of a musical instrument depends primarily on its timbre [7]. For the physics point of view, the timbre produced from an object is determined from its

© Springer Nature Switzerland AG 2020
J. Romero et al. (Eds.): EvoMUSART 2020, LNCS 12103, pp. 80–89, 2020.
https://doi.org/10.1007/978-3-030-43859-3_6

vibrational state, which characterises the object's waveform and harmonic properties. For the viewpoint of cognitive psychology [8], timbre contains complex musical features such as the attack time, the temporal envelope, the spectral envelope. Both spectral and temporal features of sounds contribute to timbre perception. For a specific musical instrument, its spectrum change is very complicated. Due to different playing techniques, the same type of musical instrument will have apparent changes in timbre. It may also be affected by mood, humidity, temperature during performance. Therefore, identifying instruments in music pieces is an exciting and challenging problem.

Generally, there are two kinds of musical data: monophonic and polyphonic music. In monophonic music, the instrument is played independently. Most of the work on instrument recognition is done under the assumption of independent performance, which simplifies the recognition task. In the case of separate recordings, musical instrument digital interface (MIDI) can store each instrument in a channel, making it easier for a single instrument to be detected. Bhalke et al. [9] proposed a musical instruments classification method based on Mel Frequency Cepstral Coefficient (MFCC) features and a Counter Propagation Neural Network. Their method obtained an accuracy of 91.84% for recognising 19 instruments. Banerjee et al. [10] proposed an approach for string instrument recognition that gave an accuracy of 89.85% using a Support Vector Machine (SVM) and 100% with a Random Forest classifier on the IRCAM dataset which contained only 4 string-family instruments.

However, music is more often polyphonic than monophonic, such as in a symphony orchestra or recording live scenes. Recognition of a single instrument in a polyphonic music recording is therefore much more difficult, and several attempts have been made for automatic recognition. Slizovskaia et al. [11] extracted instrument features through a standard bag-of-features pipeline and achieved a 67% classification accuracy on IRMAS database [12], which includes 11 different instruments in the recordings. Han et al. [6] integrated MFCC and CNN together to get a classification accuracy of 63.3% on IRMAS dataset as in [12]. Li et al. [13] achieved 82.74% accuracy on the MedleyDB dataset [14] by applying a CNN model on the raw audio data. In 2018, Hung et al. [15] achieved an 81.7% accuracy by using the constant Q-transform (CQT) and skip connection methods on MedleyDB and other datasets. In 2019, Gururani et al. [16] proposed an approach for handling weakly labeled data in an attention enable models in the OpenMIC dataset, which contains 20 instruments [17]. Their method achieved an average F1-score of approx. 81.03%.

Existing approaches for identifying instruments in monophonic music have achieved relatively satisfied performance. However, there is still a large gap for improving the accuracy of instrument recognition in polyphonic music pieces. In this paper, we propose a new framework for PMIR which uses the Hilbert-Huang Transform (HHT) to generate a feature matrix from music recordings and input these features into a new deep convolutional neural network for automatically learning instrument features from polyphonic music recordings [12]. In the experimental results, it shows that the proposed method outperforms other benchmarking methods where useful analysis and findings are carried out.

2   The Proposed Method

In the proposed framework (Fig. 1), we use the HHT [18] to generate the Hilbert spectrum for each instrument in the polyphonic music pieces. Then we build a deep convolutional neural network (DCNN) to take the Hilbert spectrum as input and produce the classification label as the output.

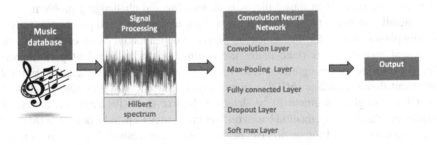

Fig. 1. The flowchart of the proposed PMIR system.

2.1   Hilbert-Huang Transform

Hilbert-Huang Transform (HHT) was proposed by Huang from NASA in 1998 [19]. This method is suitable for nonlinear and non-stationary signal analysis. At present, HHT technology has been applied in many fields, such as geophysics and biomedicine, and achieved excellent results [20]. HHT mainly includes two parts: Empirical Mode Decomposition (EMD) and Hilbert Spectral Analysis (HSA), where EMD is used to first decompose the given signals into several Intrinsic Mode Functions (IMFs). The Hilbert transform is then applied for each IMF to obtain the corresponding Hilbert Spectrum. Finally, the Hilbert spectrum of the original signal can be obtained by summing all the Hilbert spectra of the IMF. Samples of the Hilbert spectra for each instrument can be seen in Fig. 2, where they were played by cello, clarinet, flute, acoustic guitar, electric guitar, organ, piano, saxophone, trumpet, violin, and voice, respectively. It shows the performance of HHT varies according to different instruments. HHT was first used in PMIR in 2018 by Kim et al. [5], where HHT was proposed to replace the short-time Fourier Transform (STFT) and MFCC and achieved improved results. HHT and Fourier transform based calculation is given as follows for comparison:

$$x(t) = Re \sum_{i=1}^{n} a_i(t)e^{j\int w_i(t)dt} \tag{1}$$

$$x(t) = \sum_{i=1}^{\infty} a_k e^{jk(2\pi/T)t} \tag{2}$$

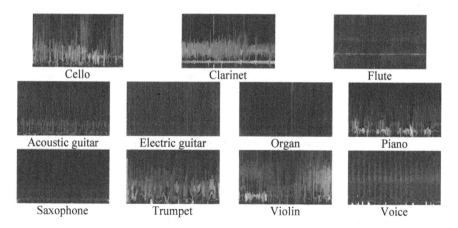

Fig. 2. Example of Hilbert spectra for each of the 11 instruments.

The main difference is that HHT can be considered as a phase shift converter whilst the traditional Fourier analysis uses a series of trigonometric basis functions for orthogonal operations on signals [21]. However, the resulting Fourier spectrum is only the weighted mean of the frequencies over a certain period and cannot be accurately described by time-frequency changes. To this end, HHT is adopted in our paper to define the instantaneous frequency for coping with more complex signals such as polyphonic music pieces.

2.2 Architecture of DCNN

Our DCNN was inspired by the VGG-16 model [22]. VGG-16 contains 16 hidden layers (13 convolutional and 3 fully connected). The pooling size is always set as 2×2 and the filter size is set as 3×3, and it shows that deepening the network layers can improve performance [22]. In our proposed DCNN model, we have 6 convolutional layers followed by a dropout layer, a fully connected layer, another dropout layer and a softmax layer. Each convolutional layer is followed by a max pooling layer. Therefore, our model has much less parameters but higher performance.

In Fig. 3, we illustrate the architecture of the proposed DCNN. The specific details of the architecture are provided in Table 1. There are six convolutional layers in the proposed model, where each convolutional layer is followed by a batch normalization layer, an activation layer, and a max pooling layer. The feature matrix produced by the Hilbert spectrum has size $135 \times 240 \times 3$, where 3 means RGB channels, and the size of each channel is 135×240. RGB channels give a colorful spectrogram. It has three channels (Red, green, blue). They can provide more information, such as frequency energy rather than grayscale. A rectified linear unit (ReLU) is selected as the activation function for each activation layer due to its popularity and ability to increase the learning speed. In the first convolutional layer the stride size is 2×2, which is changed to 1×1 for the rest of the convolutional layers. Both pool size and stride size of each max pooling layer are set to 2×2. In addition, the input for each convolutional layer is

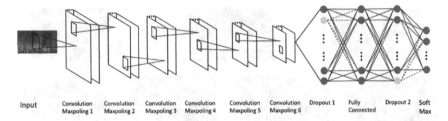

Fig. 3. Flowchart of the proposed DCNN

Table 1. Proposed DCNN structure.

Layers	Output size	Description
HHT (Input)	$135 \times 240 \times 3$	Feature matrix from Hilbert spectrum
Convolution 1	$68 \times 120 \times 32$	Filter size: 7×7; Stride size: 2×2
Max pooling 1	$34 \times 60 \times 32$	Pool size: 2×2; Stride size: 2×2
Convolution 2	$34 \times 60 \times 64$	Filter size: 5×5; Stride size: 1×1
Max pooling 2	$17 \times 30 \times 64$	Pool size: 2×2; Stride size: 2×2
Convolution 3	$17 \times 30 \times 128$	Filter size: 3×3; Stride size: 1×1
Max pooling 3	$8 \times 15 \times 128$	Pool size: 2×2; Stride size: 2×2
Convolution 4	$8 \times 15 \times 256$	Filter size: 3×3; Stride size: 1×1
Max pooling 4	$4 \times 7 \times 256$	Pool size: 2×2; Stride size: 2×2
Convolution 5	$4 \times 7 \times 512$	Filter size: 3×3; Stride size: 1×1
Max pooling 5	$2 \times 3 \times 512$	Pool size: 2×2; Stride size: 2×2
Convolution 6	$2 \times 3 \times 1024$	Filter size: 3×3; Stride size: 1×1
Max pooling 6	$1 \times 1 \times 1024$	Pool size: 2×2; Stride size:2×2
Dropout 1	1024	Dropout fact: 0.25
Fully connected	1024	Output size: 11
Dropout 2	1024	Dropout fact: 0.25
Softmax	11	Softmax function

same-padded in order to preserve the spatial resolution. The number of filters for each convolution layer is twice that of the previous layer, increasing from 32 up to 1024 at the last layer. After the final max pooling layer, a dropout layer is added before and after the fully connected layer to avoid overfitting. In the end, the Softmax function is used for 11 instruments classification.

3 Experiments

In this experiment, 6705 audio files were extracted from the IRMAS dataset [12] and used to investigate the performance of the proposed method. The audio files are in 16-bit stereo .wav format, sampled at 44.1 kHz, and have a length of 3 s. There are 11 different instruments in the music (see Fig. 2). The percentage of training, validation, and testing data was set to 55%, 15% and 30% respectively.

3.1 Key Parameters

Adjusting hyper-parameters to improve network performance is a vital step in deep learning. To get the optimal performance, we only vary the minibatch size from 50 to 300 in intervals of 50. From Table 2, the best testing accuracy can achieve 85% when the minibatch size is 250. As a result, the batch size in our experiments was set to 250. With minibatch size equaling 250, the classification accuracy of proposed model with varying dropout factor from 0 to 0.5 at a step of 0.05 is shown in Fig. 4. Based on the results, the dropout rate is set as 0.25 in our work. For the other parameters, Adam is used as the optimiser with a learning rate of 0.01.

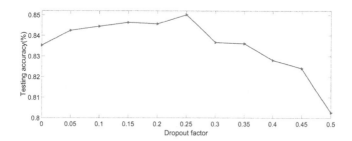

Fig. 4. Classification performance with different dropout factors.

Table 2. Training performance with different minibatch size

Batch size	50	100	150	200	250	300
Testing accuracy (%)	84.30	84.96	84.09	84.54	**85.01**	84.67

3.2 Comparison with Other Methods

To further evaluate the efficacy of the proposed PMIR framework, three conventional approaches are used for benchmarking in terms of precision, recall and F1-measurement [23]. These conventional frameworks are based on Audio Content Analysis (ACA) system [20, 24] and three machine learning model, i.e. random forest (RF) [4], SVM [1] and shallow neural network (SNN) [2]. These conventional methods are representative in the field of machine learning, and the ACA features include most of the features used in music analysis. The tree number of RF is 300, the LIBSVM toolbox [1] is selected as SVM learner, and the neuron number in the hidden layer in SNN is set as 70. For each music piece, the ACA system is used to extract 19 features from both time and frequency domains such as Peak envelop, autocorrelation coefficients, MFCCs, and pitch chroma. Then these features will concentrate into a feature vector with the length of 105 and be entered into the machine learning model for instruments' classification. What's more, we compared it with a new HAS-IMF CNN algorithm proposed in 2018 [5], using the same dataset and computing environment.

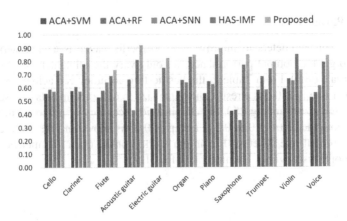

Fig. 5. F1-measurement of each instrument for five methods.

Table 3. Overall precision, recall and F1-measurement of five methods.

Methods	Precision	Recall	F1
ACA+SVM [1]	0.53	0.54	0.53
ACA+SNN [2]	0.57	0.56	0.56
ACA+RF [4]	0.61	0.62	0.61
HAS-IMF [5]	0.77	0.80	0.78
Proposed	**0.82**	**0.85**	**0.84**

Experimental results including the overall precision, recall and F1-measurement are presented in Table 3, and the F1-measurement of each instrument is shown in Fig. 5. As can be seen, the proposed PMIR framework generates the best overall performance and outperforms conventional frameworks in classifying individual instruments. In Table 3, the ACA features are combined in the first three classifiers, i.e. SVM, SNN, and RF. Among them, the RF based method seems the best, yet it is significantly poorer than the HAS-IMF, due mainly to the integration of the HHT spectrogram and CNN model. However, thanks to our improved CNN structure, the proposed model has significantly outperformed all others including HAS-IMF, where the precision, recall, and F1 measures are improved by 5%, 5% and 6%, respectively.

In our model, the batch normalization and Max pooling are followed by each convolutional layer, and the dropout layers are put before and after the fully connected layer. However, in HAS-IMF, a dropout layer is followed by every two convolutional layers. Although the dropout layer can reduce the training time, too many dropout layers may lead to the network not being fully trained. Furthermore, it does not include a batch normalization layer which may lead to the data becoming unbalanced. Therefore, our proposed CNN model outperforms HAS-IMF by 6%.

In Figs. 5 and 6, the classification performance of individual instruments is presented. As can be seen HAS-IMF and the proposed method are significantly better than the fusion of traditional features (ACA) and other machine learning techniques. For the

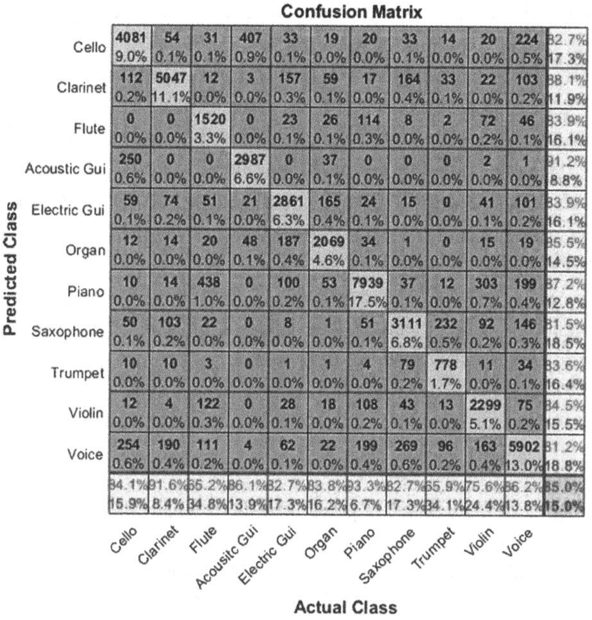

Confusion Matrix

Predicted Class \ Actual Class	Cello	Clarinet	Flute	Acousitc Gui	Electric Gui	Organ	Piano	Saxophone	Trumpet	Violin	Voice	
Cello	4081 / 9.0%	54 / 0.1%	31 / 0.1%	407 / 0.9%	33 / 0.1%	19 / 0.0%	20 / 0.0%	33 / 0.1%	14 / 0.0%	20 / 0.0%	224 / 0.5%	82.7% / 17.3%
Clarinet	112 / 0.2%	5047 / 11.1%	12 / 0.0%	3 / 0.0%	157 / 0.3%	59 / 0.1%	17 / 0.0%	164 / 0.4%	33 / 0.1%	22 / 0.0%	103 / 0.2%	88.1% / 11.9%
Flute	0 / 0.0%	0 / 0.0%	1520 / 3.3%	0 / 0.0%	23 / 0.1%	26 / 0.1%	114 / 0.3%	8 / 0.0%	2 / 0.0%	72 / 0.2%	46 / 0.1%	83.9% / 16.1%
Acoustic Gui	250 / 0.6%	0 / 0.0%	0 / 0.0%	2987 / 6.6%	0 / 0.0%	37 / 0.1%	0 / 0.0%	0 / 0.0%	0 / 0.0%	2 / 0.0%	1 / 0.0%	91.2% / 8.8%
Electric Gui	59 / 0.1%	74 / 0.2%	51 / 0.1%	21 / 0.0%	2861 / 6.3%	165 / 0.4%	24 / 0.1%	15 / 0.0%	0 / 0.0%	41 / 0.1%	101 / 0.2%	83.9% / 16.1%
Organ	12 / 0.0%	14 / 0.0%	20 / 0.0%	48 / 0.1%	187 / 0.4%	2069 / 4.6%	34 / 0.1%	1 / 0.0%	0 / 0.0%	15 / 0.0%	19 / 0.0%	85.5% / 14.5%
Piano	10 / 0.0%	14 / 0.0%	438 / 1.0%	0 / 0.0%	100 / 0.2%	53 / 0.1%	7939 / 17.5%	37 / 0.1%	12 / 0.0%	303 / 0.7%	199 / 0.4%	87.2% / 12.8%
Saxophone	50 / 0.1%	103 / 0.2%	22 / 0.0%	0 / 0.0%	8 / 0.0%	1 / 0.0%	51 / 0.1%	3111 / 6.8%	232 / 0.5%	92 / 0.2%	146 / 0.3%	81.5% / 18.5%
Trumpet	10 / 0.0%	10 / 0.0%	3 / 0.0%	0 / 0.0%	1 / 0.0%	1 / 0.0%	4 / 0.0%	79 / 0.2%	778 / 1.7%	11 / 0.0%	34 / 0.1%	83.6% / 16.4%
Violin	12 / 0.0%	4 / 0.0%	122 / 0.3%	0 / 0.0%	28 / 0.1%	18 / 0.0%	108 / 0.2%	43 / 0.1%	13 / 0.0%	2299 / 5.1%	75 / 0.2%	84.5% / 15.5%
Voice	254 / 0.6%	190 / 0.4%	111 / 0.2%	4 / 0.0%	62 / 0.1%	22 / 0.0%	199 / 0.4%	269 / 0.6%	96 / 0.2%	163 / 0.4%	5902 / 13.0%	81.2% / 18.8%
	84.1% / 15.9%	91.6% / 8.4%	65.2% / 34.8%	86.1% / 13.9%	82.7% / 17.3%	83.8% / 16.2%	93.3% / 6.7%	82.7% / 17.3%	55.9% / 44.1%	75.6% / 24.4%	86.2% / 13.8%	85.0% / 15.0%

Fig. 6. Confusion matrix of proposed methods in 11 instruments.

violin, HAS-IMF gives the best result. But with a better structure of deep learning network, our proposed method produces better performance on the rest of the individual instruments. Another finding is that the performance of individual instruments is potentially related to the types of instruments. For example, as can be seen in Fig. 6, the saxophone sometimes is misclassified to a clarinet and trumpet, because both saxophone and clarinet belong to the woodwind family, and both saxophone and trumpet belong to the wind family. In addition, piano is mostly misclassified into violin and flute. The main reason is that this dataset has many variables, it cannot only discuss about timbre. The pitch of violin and flute is very high, and the piano sometimes composes the main melody by a high pitch. Therefore, there is the confusion of the piano with violin and flute. Also, the human voice is often misclassified into cello, clarinet, piano and saxophone since it is very complicated, and its pitch or timbre may be close to some of the instruments.

4 Conclusions

In this paper, a new framework for predominant musical instrument recognition (PMIR) in polyphonic music was proposed. First the Hilbert-Huang Transform (HHT) is used to generate the Hilbert spectrum of the audio data, which is then used as input to a deep convolutional neural network (DCNN). The optimal DCNN model is trained on the IRMAS dataset, and objective evaluation has shown that the proposed method can give a classification accuracy of 85%. The proposed method also outperforms three conventional approaches, which shows that the image-based deep learning method has good potential for music instrument recognition. In the future, the MedleyDB dataset [14] and OpenMIC-2018 dataset [17] will be used to evaluate the proposed method and more comprehensive experiments will be done to further validate the usefulness and effectiveness of the proposed method. Key parameters' selection and network structure will also be optimised. In addition, the performance of state-of-art deep learning models such as stacked autoencoders [25], and residual neural networks [26] on PMIR will be investigated. Optimized model like gravitational search algorithm [27] can be applied to improve the feature selection.

Acknowledgements. The authors would like to thank, Dr Yijun Yan for useful discussions and Calum MacLellan for kindly proofreading of the paper.

References

1. Chang, C.C., Lin, C.J.: LIBSVM: a library for support vector machines. ACM Trans. Intell. Syst. Technol. **2**, 1–27 (2011)
2. Battiti, R.: First-and second-order methods for learning: between steepest descent and Newton's method. Neural Comput. **4**, 141–166 (1992)
3. Downie, J.S., Ehmann, A.F., Bay, M., Jones, M.C.: The music information retrieval evaluation eXchange: some observations and insights. Stud. Comput. Intell. **274**, 93–115 (2010)
4. Liaw, A., Wiener, M.: Classification and regression by randomForest. R News **2**(3), 18–22 (2002)
5. Kim, D., Sung, T.T., Cho, S., Lee, G., Sohn, C.-B.: A single predominant instrument recognition of polyphonic music using CNN-based timbre analysis. Int. J. Eng. Technol. **7**, 590–593 (2018)
6. Han, Y., Kim, J., Lee, K., Han, Y., Kim, J., Lee, K.: Deep convolutional neural networks for predominant instrument recognition in polyphonic music. IEEE/ACM Trans. Audio Speech Lang. Process. (TASLP) **25**, 208–221 (2017)
7. Fletcher, N.H., Rossing, T.D.: The Physics of Musical Instruments. Springer, Heidelberg (2012)
8. McAdams, S., Giordano, B.L.: The perception of musical timbre. In: The Oxford Handbook of Music Psychology, pp. 113–123 (2016)
9. Bhalke, D., Rao, C.R., Bormane, D.S.: Automatic musical instrument classification using fractional fourier transform based-MFCC features and counter propagation neural network. J. Intell. Inf. Syst. **46**, 425–446 (2016)
10. Banerjee, A., Ghosh, A., Palit, S., Ballester, M.A.F.: A novel approach to string instrument recognition. In: Mansouri, A., El Moataz, A., Nouboud, F., Mammass, D. (eds.) ICISP 2018. LNCS, vol. 10884, pp. 165–175. Springer, Cham (2018). https://doi.org/10.1007/978-3-319-94211-7_19
11. Slizovskaia, O., Gómez, E., Haro, G.: Automatic musical instrument recognition in audiovisual recordings by combining image and audio classification strategies. In: SMC 2016 – 13th Sound and Music Computing Conference, Proceedings, pp. 442–447 (2016)
12. Bosch, J.J., Janer, J., Fuhrmann, F., Herrera, P.: A comparison of sound segregation techniques for predominant instrument recognition in musical audio signals. In: ISMIR, pp. 559–564 (2012)

13. Li, P., Qian, J., Wang, T.: Automatic instrument recognition in polyphonic music using convolutional neural networks. arXiv preprint: arXiv:1511.05520 (2015)
14. Bittner, R.M., Salamon, J., Tierney, M., Mauch, M., Cannam, C., Bello, J.P.: MedleyDB: a multitrack dataset for annotation-intensive mir research. In: ISMIR, pp. 155–160 (2014)
15. Hung, Y.N., Chen, Y.A., Yang, Y.H.: Multitask learning for frame-level instrument recognition. In: Proceedings of the IEEE International Conference on Acoustics, Speech and Signal Processing, pp. 381–385 (2019)
16. Gururani, S., Sharma, M., Lerch, A.: An Attention Mechanism for Musical Instrument Recognition. arXiv preprint: arXiv:1907.04294 (2019)
17. Humphrey, E., Durand, S., McFee, B.: OpenMIC-2018: an open data-set for multiple instrument recognition. In: ISMIR, pp. 438–444 (2018)
18. Sandoval, S., De Leon, P.L., Liss, J.M.: Hilbert spectral analysis of vowels using intrinsic mode functions. In: 2015 IEEE Workshop on Automatic Speech Recognition and Understanding (ASRU), pp. 569–575. IEEE (2015)
19. Müller, M.: Fundamentals of Music Processing: Audio, Analysis, Algorithms. Applications. Springer, Cham (2015). https://doi.org/10.1007/978-3-319-21945-5
20. Lerch, A.: An Introduction to Audio Content Analysis: Applications in Signal Processing and Music Informatics. Wiley-IEEE Press, Hoboken (2012)
21. Ayenu-Prah, A., Attoh-Okine, N.: Comparative study of Hilbert-Huang transform, Fourier transform and wavelet transform in pavement profile analysis. Veh. Syst. Dyn. **47**, 437–456 (2009)
22. Simonyan, K., Zisserman, A.: Very deep convolutional networks for large-scale image recognition. In: 3rd International Conference on Learning Representations, ICLR 2015 - Conference Track Proceedings (2014)
23. Yan, Y., et al.: Cognitive fusion of thermal and visible imagery for effective detection and tracking of pedestrians in videos. Cogn. Comput. **10**, 94–104 (2018)
24. Peeters, G.: A large set of audio features for sound description (similarity and classification). CUIDADO project IRCAM technical report (2004)
25. Zabalza, J., et al.: Novel segmented stacked autoencoder for effective dimensionality reduction and feature extraction in hyperspectral imaging. Neurocomputing **185**, 1–10 (2016)
26. He, K., Zhang, X., Ren, S., Sun, J.: Deep residual learning for image recognition. In: Proceedings of the IEEE Conference on Computer Vision and Pattern Recognition, pp. 770–778 (2016)
27. Sun, G., Ma, P., Ren, J., Zhang, A., Jia, X.: A stability constrained adaptive alpha for gravitational search algorithm. Knowl.-Based Syst. **139**, 200–213 (2018)

Genetic Reverb: Synthesizing Artificial Reverberant Fields via Genetic Algorithms

Edward Ly$^{(\boxtimes)}$(iD) and Julián Villegas(iD)

University of Aizu, Aizu-Wakamatsu, Fukushima 965-8580, Japan
{m5222120,julian}@u-aizu.ac.jp
http://onkyo.u-aizu.ac.jp/

Abstract. We present Genetic Reverb, a user-friendly VST 2 audio effect plugin that performs convolution with Room Impulse Responses (RIRs) generated via a Genetic Algorithm (GA). The parameters of the plugin include some of the standard room acoustics parameters mapped to perceptual correlates (decay time, intimacy, clarity, warmth, among others). These parameters provide the user with some control over the resulting RIRs as they determine the fitness values of potential RIRs. In the GA, these RIRs are initially generated via a Gaussian noise method, and then evolved via truncation selection, multi-point crossover, zero-value mutation, and Gaussian mutation. These operations repeat until a certain number of generations has passed or the fitness value reaches a threshold. Either way, the best-fit RIR is returned. The user can also generate two different RIRs simultaneously, and assign each of them to the left and right stereo channels for a binaural reverberation effect. With Genetic Reverb, the user can generate and store new RIRs that represent virtual rooms, some of which may even be impossible to replicate in the physical world. An original musical composition using the Genetic Reverb plugin is presented to demonstrate its applications. (The source code and link to the demo track is available at https://github.com/edward-ly/GeneticReverb).

Keywords: Convolution reverb · Genetic algorithms · Impulse responses · Room acoustics · Signal processing

1   Introduction

In most common situations, sound is perceived within enclosures such as rooms, venues, caves, etc. These enclosures imprint their characteristics (i.e., their transfer functions) in the sound depending on where the sound source and the listener are within the room. The direct sound travels the shortest path between a source and a listener, other paths may include several reflections off the boundaries of the enclosure. Hence, reflected sounds arrive later than the direct sound and are in general weaker because of the energy absorption of the reflecting surfaces as

© Springer Nature Switzerland AG 2020
J. Romero et al. (Eds.): EvoMUSART 2020, LNCS 12103, pp. 90–103, 2020.
https://doi.org/10.1007/978-3-030-43859-3_7

well as the transmission medium. The collection of early and late reflections are usually referred as reverberation, and although it is somewhat detrimental to speech intelligibility [5], reverberation is sought after in music as an expressive tool [11].

The most direct form of including reverberation in audio recordings is to capture the latter in spaces with desired characteristics. Alternatively, one can record the room characteristics (i.e., its impulse response—IR of the room) and filter audio with them.

Recording the Impulse Responses (IRs) of different venues is expensive and not always possible, so other methods to add artificial reverberation can be used instead without apparent quality detriment. Delay lines, comb filters, feedback delay networks, etc. are some of the most common techniques to create artificial reverberation. An exhaustive description of these methods is beyond the scope of this article, but the interested reader is referred to [12].

Regardless of the relatively large amount of alternatives for adding reverberation, uniqueness is a characteristic often sought in the artistic process. In our research, we introduce yet another method to create a room IR using evolutive techniques so that the resulting IR is most likely unique while satisfying some constraints imposed by the user.

The rest of this article is organized as follows: Sect. 2 explains the image method to generate the initial IRs used in our solution. Section 3 delves into the Genetic Algorithm (GA) implemented to modify the initial IR. Section 4 introduces the plugin technology used to deploy our solution as a real-time audio effect. Section 5 presents a brief evaluation followed by concluding remarks.

2   Background

The idea of applying evolutive techniques to real-time digital signal processing is not new, but compared to other techniques in artificial intelligence applied to DSP, there seems to be little research and few actual implementations. One example comes from Macret & Pasquier [6], who generated sound synthesizers in Pure-Data (Pd) via Mixed-Typed Cartesian Genetic Programming (MT-CGP). Pure-Data is an open-source visual programming language for DSP synthesis, with a graphical interface that visualizes each Pd program, or "patch", as a directed graph structure. Genetic Programming (GP) is a technique similar to Genetic Algorithms (GAs) except that entire graph structures (the genome for these Pd patches) are subjected to evolution. It is a useful tool for generating new sounds and instruments that closely resemble other target sounds, but the average end user may not be familiar with Mel-Frequency Cepstral Coefficients (MFCC), which are used in the GUI as the input parameters for the MT-CGP algorithm to represent the target sound.

Collins [2] goes further by providing a library built in the SuperCollider programming language that can apply GAs to both sound synthesis and real-time signal processing, including filtering and reverberation. Using SuperCollider's built-in Graphical User Interface GUI and programming tools, a framework for other developers to build software applications using GAs was provided. Collins does not mention any ability for a GA to evolve a population of IRs for convolution reverberation, but rather other devices for other approaches including multi-tap delay lines, feedback delay networks, and Schroeder reverberators (these can be thought as a series of allpass and comb filters). In addition, currently (as of Jan. 2020), the portability of SuperCollider code to Virtual Studio Technology (VST) plugins (a popular cross-platform audio software interface widely used in music production) is very limited at best. While a SuperColliderAU wrapper is available to compile SuperCollider code as an Audio Units (AU) plugin, this competing technology is only compatible with the Mac operating system environment.

2.1   Impulse Response of a Box-Shaped Room

An image method for simulating the reverberation of a box-shaped room was originally proposed by Allen & Berkley [1]. In their method, a room was projected onto a three-dimensional space to determine which sound reflections contributed to an IR. The same projection was also used to calculate the effective distance and associated gain for each contributing reflection. As a metaphorical example, imagine a listener and a sound source located arbitrarily within a room with mirrors on all sides. Visually, perfect mirrors will create an infinite number of images of the object outside the boundaries of the room. The listener could see the same sound source reflected an infinite number of times. Assuming that these images emitted sound, each sound reflection will be delayed by a time equal to the distance between the listener and each object image divided by the speed of sound. Additionally, each reflection will lose power depending on the number of "walls" the sound has to "pass through" before reaching the listener's location. Practically, reflections greater than a certain order (or the number of walls between the listener and a reflected sound source) are ignored in software implementations for being too quiet to be considered.

Many extensions and applications based on the Allen & Berkley method would later be proposed. Several implementations of the original method into modern programming languages have been developed as well, with Habets [3] providing one such implementation in the Matlab programming language [7]. The running time of the original method was deemed to be too long to be feasibly useful for real-time signal processing (at least in our tests), especially when multiple IRs have to be generated to evolve an entire population of them. Alternatively, more recent revisions to the image method that tackled the said performance issues have also been proposed, such as those by Kristiansen et al. [4] and McGovern [8]. Kristiansen et al. uses the image method to extend the length of an existing IR, using known lower-level reflections to calculate higher-level ones, while McGovern [8] improves on the original image method algorithm by using look-up tables and sorting to prevent unnecessary calculations, greatly reducing computation time. Initially, we implemented the fast image method proposed by McGovern as the baseline model for generating our initial population. However, since our GA ultimately outputs IRs that are not necessarily modeled after box-shaped rooms regardless of the nature of the initial population, we came to realize that there was no reason why we could not use other non-box shaped models in the first place. In addition, a number of our output IRs sounded erratic and fragmented due to the crossover of multiple distinct IRs, which may be an unwanted characteristic of a feasible output IR. The following section describes in detail our current implementation.

3   Implementation

We implemented a classic GA on a population of room IRs where the genome for each IR individual consists of an array of real-valued genes in the range of $[-1, 1]$ in the time domain. Each non-zero gene represents the arrival of a sound reflection at a certain point in time, with the first gene assumed to be non-zero and represent the direct sound, or the start of the IR. Our GA applies multiple rounds (or "generations") of selection, crossover, and mutation, which will remove reflections, add new reflections, and change the amplitude of existing reflections at random until a created IR closely adheres to the user constraints.

Since a typical end user may not understand what a GA does, we do not want the standard GA parameters, such as the maximum number of generations to execute or the mutation rate, to be constraints that can be controlled by the user. Instead, we have implemented some of the standard room acoustics parameters as not only additional parameters for the GA, but also constraints that the user can control and understand. The acoustics parameters that we have implemented so far in our plugin are summarized in Table 1. Except for early decay time (EDT), which is usually equal to one-sixth of the T_{60} reverberation time in a typical IR, these parameters can be controlled by the user in order to control the overall shape of the IR produced by the GA.

Table 1. List of acoustics parameters used in the plugin along with their colloquial names and definitions.

Parameter	Common name	Definition
Reverberation Time (T_{60})	Decay Time	Time for the impulse response to decay 60 dB from the initial amplitude
Early Decay Time (EDT)	Early Decay Time, Reverberance	Time for the impulse response to decay 10 dB from the initial amplitude
Initial Time Delay Gap (ITDG)	Intimacy	Time between the arrival of the initial sound and the arrival of the next reflected sound
Clarity (C_{80})	Clarity	Ratio in dB between energy levels of early reflections (<80 ms after initial reflection) and late reflections (>80 ms after initial reflection)
Bass Ratio (BR)	Warmth	Ratio in dB between low-frequency (125–500 Hz) to mid-frequency (0.5–2.0 kHz) content

3.1 Initialization

In the initialization step, we generated a population of room IRs using a Gaussian noise method. We started with an array of normally distributed random numbers ($\mu = 0$, $\sigma = 0.2$) for the duration of the desired IR. The length of the array, or the number of samples in the IR, is determined by the reverberation time T_{60} of the desired IR along with the sampling rate of the IR to generate. In our plugin, the sampling rate is set to a constant 16 kHz in order to minimize computation time.

Then, we randomly removed certain samples in the array, zeroing them out so that the remaining samples serve as the reflections in an IR individual. The probability that a reflection occurs at time t (in seconds) is

$$P(t) = \begin{cases} 1, & t = 0 \\ 0, & 0 < t < t_g \\ 1, & t = t_g \\ 1 - p^t, & t > t_g \end{cases} \tag{1}$$

for some probability constant $p \in (0, 1)$ and an Initial Time Delay Gap (ITDG) t_g of the desired IR. This model allows us to simulate typical IRs where the reflection density (i.e., number of reflections per second) of early reflections is smaller compared to later reflections, as well as control the ITDG of the IR with exact accuracy. However, we found that only p values ranging from $\sim$0.2 to $\sim$0.7 provide a reflection density similar to that of a typical IR obtained through the previous image method. Thus, the p value of each individual in the initial population is a random value in this range.

Afterwards, we imposed a gain factor given by $G(t) = d^{(t/T_{60})}$ for a decay constant $d \in (0, 1)$. This changes the amplitudes of each of the IR reflections so that they tend to decay exponentially over time depending on the decay constant d and the T_{60} value of the desired IR. However, we also found that $\sim 2 \times 10^{-4} \leq d \leq \sim 4 \times 10^{-4}$ produce small enough gains at the ends of the IRs so that an individual IR's T_{60} is similar to the one specified by the user. Thus, the decay value d of each individual in the initial population is a random value in this range as well.

Finally, we forced the amplitude of the initial reflection of each IR in the entire population to unity. This gives our IRs some consistency in terms of the amplitude of the arrival of the direct sound as well as the overall intensity of the IRs. With all of these restrictions in place, an initial population of IRs was seeded and used as a starting point from which other parameters can be sought.

3.2 Fitness

Each IR in a population is assigned a fitness value based on how closely it matches user specifications. In our implementation, the fitness function of the GA assigns an "error" or "loss" value, with figures toward zero representing a better "fit". This loss is computed as the sum of the absolute values of the z-scores of the standard room acoustics parameters summarized in Table 1 (except ITDG, since it can be controlled with exact accuracy).

It is possible to compute these parameters from the generated IRs and then compare them with the desired values to find the closest match. To find T_{60}, we first calculated the IR's Schroeder curve $S(t)$ (a.k.a. energy decay curve) [10], formally defined as the tail integral of the square of an IR $h(\tau)$ at time $\tau = t$:

$$S(t) = \int_t^\infty h^2(\tau) \, d\tau. \tag{2}$$

Expressing $S(t)$ in dB, it is possible to locate the instances where the sound energy decays 5 dB and 35 dB from the initial amount, this time difference is known as T_{30}. Doubling T_{30} yields an accurate approximation of T_{60}.

Early Decay Time (EDT) is computed in a similar manner, except that the time difference is computed between the arrival of the direct sound (0 dB) and the time when the total energy decays 10 dB.

Clarity (C_{80}) is determined using Schroeder curves as well. C_{80} is defined in terms of $S(t)$ as

$$C_{80} = 10 \log \left(\frac{S(0) - S(0.08)}{S(0.08)} \right) \text{dB}, \tag{3}$$

where $t = 0$ refers to the time of arrival of the direct sound.

For bass ratio, we transformed the IR into the frequency domain via the Fast Fourier Transform (FFT), then computed the sound energy ratio (in dB) between the 125–500 Hz and 0.5–2.0 kHz. Bass ratio is formally defined as the ratio of EDTs rather than total sound energy in the same frequency bins [9], but the implemented method is faster with no apparent detriment in accuracy. In addition, expressing bass ratio in dB units (as opposed to a unit-less value) makes the parameter easier to understand for the end user.

Finally, the z-scores of the generated IR parameters were computed, using the desired IR parameter values as the mean for each z-score. The error value of each IR is then determined by the sum of the absolute values of all the z-scores (i.e., the total number of standard deviations away from the optimal solution). Because of the differences in units and in numerical scale between the IR parameters, this error value allows us to weigh the individual parameters equally, and is useful not just as a standard statistical metric, but also for removing outlier IRs.

3.3 Genetic Operations

In each generation, the IR population undergoes three successive operations (i.e., selection, crossover, and mutation) in order to search for a better match. We decided to implement truncation selection and three-point crossover, as well as zero-value mutation and Gaussian mutation in our GA.

Truncation selection is one of the simplest selection methods: the bottom percentile of IR individuals (those with the worst fit) are removed from the pool each generation. The removed IRs are replaced with new ones through a three-point crossover operation: Given two random parent IRs $h_1(t)$ and $h_2(t)$, and three random distinct points in time τ_1, τ_2, and τ_3 ($\tau_1 < \tau_2 < \tau_3$) within the length of the parent IRs, the child IR $h_{1,2}(t)$ would be defined as

$$h_{1,2}(t) = \begin{cases} h_1(t), & t < \tau_1 \\ h_2(t), & \tau_1 \leq t < \tau_2 \\ h_1(t), & \tau_2 \leq t < \tau_3 \\ h_2(t), & t \geq \tau_3 \end{cases}. \tag{4}$$

Crossover with an arbitrary number of points follows a similar pattern, alternating parents after each point.

Finally, we perform two rounds of mutation operations, where each sample has a very small probability ($<1\%$) of changing its value completely. First, in zero-value mutation, we take a random set of samples and set the values of these samples to zero, deleting any reflections that occur at those points in time. Then, in Gaussian mutation, we take another random set of samples and add to each sample a unique, random number from the normal distribution ($\mu = 0, \sigma = 0.1$). This produces one of two effects: either a new reflection that has not existed before is created, or the amplitude of an already existing reflection is changed. Only samples that occur after the initial ITDG time are mutated in order to preserve the ITDG of the IR.

3.4   Termination

After each round of genetic operations, the fitness of the new IR population is re-calculated to start a new iteration in the evolution cycle, but this is also when the GA determines whether or not it should continue modifying the population, or stop and return the best-fit IR found so far. This decision is based on three terminating conditions:

1. The fitness value of the best IR found is below a certain threshold value,
2. The fitness value of the best IR found does not decrease after some number of generations ("plateau length"), or
3. A predetermined limit on the number of generations to execute has been reached.

The fitness threshold, the plateau length, and the maximum number of generations are all parameters controlled by the plugin to limit computation time. The GA stops whenever at least one of these terminating conditions is met.

4   Matlab Audio Plugin

In order to develop a working prototype, we implemented Genetic Reverb as a subclass of Matlab's built-in audioPlugin and System classes. These classes can be compiled into a VST 2 plugin compatible with both Windows and Mac

Table 2. IR-parameters used in the plugin with their valid range and descriptions

Parameter	Valid range	Description
Decay time	$[0.4, 10]$ s	T_{60} decay time of the desired IR
Intimacy	$[1, 100]$ ms	ITDG of the desired IR
Clarity	$[-5, 5]$ dB	Amount of clarity in the desired IR
Warmth	$[-5, 5]$ dB	Amount of warmth in the desired IR
Predelay (L)	$[0.5, 200]$ ms	Amount of time to delay before the arrival of the initial sound in the left channel IR
Predelay (R)	$[0.5, 200]$ ms	Amount of time to delay before the arrival of the initial sound in the right channel IR
Quality	{"Low", "Mid", "High"}	Sets various parameters for the GA (see Table 3)
Mono/stereo	{"Mono", "Stereo"}	Either generate one IR for both channels (mono) or generate a different IR for each channel (stereo)
Dry/wet	$[0, 100]$%	Balance between the dry input signal (0%) and the processed one (100%)
Output gain	$[-60, 12]$ dB	Gain of the mixed dry/wet signal
Generate room	{true, false}	Toggling this switch generates a new IR using the current parameters
Toggle to save	{true, false}	Toggling this switch saves the current IR as a binary file in the plugin directory

operating systems, and let us define, among other things, parameters that a user can set, reactions to changes in these parameters, as well as the DSP needed to perform tasks in real-time. Table 2 summarizes these parameters, the valid range of values, and a brief description of each, while Table 3 lists the current mapping between the "Quality" parameter and the GA parameters used for the plugin. The "Quality" parameter of the plugin was implemented so that the user has some control over the balance between the computation time of the GA and the fitness value of the output IR(s).

Matlab's toolboxes and APIs also allowed us to implement some additional features that are useful for end users in a typical production environment as well as additional processing on the IRs generated by our GA. For real-time convolution reverb, the plugin always contains two active IRs that are convolved with the input signal at any given time, one for each stereo channel. Each time the "Generate Room" switch is toggled, it triggers the GA to generate two new IRs (only one in mono mode that is copied for both channels to use). Because the differences between the two IRs may be large, especially in terms of overall gain, the IRs are normalized so their Root Mean Square (RMS) amplitudes are equal. Then, a small amount of predelay (specified by the "Predelay" parameters) is applied to the IRs before they are sent to be convolved with the input signal.

Table 3. List of parameter values used in the genetic algorithm within the plugin depending on reverb quality. Fitness threshold and mutation probability were constant at 0.1 and 0.001, respectively.

Parameter	Quality		
	Low	Mid	High
Population size	10	20	20
Selection size	4	8	8
Max. number of generations	10	10	20
Plateau length	3	3	5

For the convolution itself, Matlab provides a DSP system object that can perform partitioned convolution on the input signal with our IRs via FFT-based Finite Impulse Response (FIR) filtering. We computed the convolution in blocks of 1024 samples at a time as a starting point, balancing between latency and CPU usage while still making convolution reverb possible to execute in real-time (as opposed to non-partitioned convolution). Figure 1 illustrates the basic flow of data and audio within the plugin, while Fig. 2 shows the current user interface of the plugin.

5   Evaluation

To assess how closely the characteristics of the generated IRs match the user parameters, 1000 IRs were generated using all of the same internal parameters used by the plugin. The parameter values for intimacy, clarity, and warmth were randomly selected within the valid range for each IR, while early decay time is set to one-sixth of the T_{60} decay time. Meanwhile, T_{60} was restricted to $t \in \{0.625, 1.25, 2.5, 5\}$ s, 250 IRs each time, in order to measure the relationship between T_{60} decay time and computation time. This process was repeated three times, one for each of the possible "quality" settings. Finally, we used Matlab's built-in time-keeping tools to record the execution times of the GA in addition to recording the fitness values of each IR and the individual error values of each fitness parameter. Tables 4, 5, and 6 provide a summary of the results, reporting the mean and standard deviation for each group of quality setting and T_{60}. Tests were run on a Windows 10 Home laptop computer equipped with an Intel Core i9-8950HK notebook processor running Matlab version R2019b Update 3.

The relatively high fitness values can be partially attributed to the fact that certain combinations of parameter values would result in IRs that would be very difficult to produce even for our GA (e.g., IRs with long T_{60} reverberation times and high C_{80} clarity values). Otherwise, if reasonable values are chosen for these parameters, then IRs that closely match the desired parameter values can be obtained through our GA. Table 7 reports the mean and standard deviation of run time, fitness value, and parameter error amounts of 250 IRs obtained through our GA using the initial settings of the plugin. Currently, these include

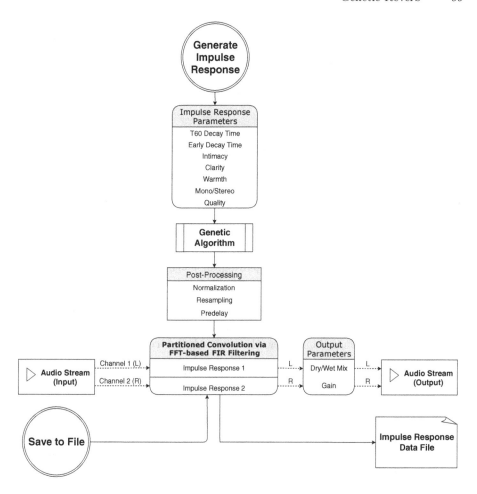

Fig. 1. Flow chart illustrating the functionality of the plugin.

a T_{60} decay time of 1 s, an ITDG of 10 ms, C_{80} set to +3 dB, BR set to 0 dB, and "low" quality GA settings.

5.1 Limitations

While Matlab's Audio Toolbox and DSP System Toolbox APIs allow for the prototyping and generation of VST plugins with relative ease, there are a few major hurdles that hinders a full implementation in some cases: first, there are a limited number of data types for parameters meaning that some GUI elements such as buttons cannot be directly implemented. For us, this is more of an annoyance than a setback since a toggle switch can act as a valid substitute for triggering actions such as generating an IR or saving a file, but this interface is not as intuitive as it could be for an end user.

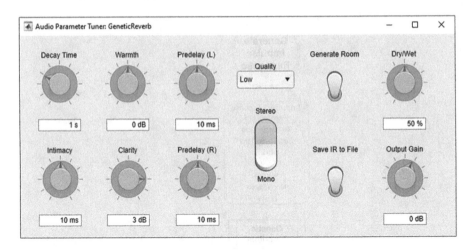

Fig. 2. Program window displaying the plugin's user interface.

Another feature that would be required for a plugin like this to be feasible in a production environment is the ability to save desirable IRs generated by the plugin in order to reuse them later. However, Matlab code generation prevents us from saving IRs directly as WAV audio files (or in most other audio formats), so binary files are used instead. These can be converted into WAV audio files using an external program such as a separate Matlab script, which we implemented and provided along with the plugin. Regardless, users must be careful to confirm that the plugin has write-access to their specified directory as well.

The DSP system object that Matlab uses to perform convolution reverb places a restriction in that the length of the array containing the filter coefficients cannot be changed dynamically. This means that the number of samples in the IR cannot be changed neither. A possible workaround that we implemented is to create multiple FIR arrays with varying filter lengths. Then, we can use the

Table 4. Evaluation of the GA using low quality settings, reporting the mean and standard deviation (in parentheses) of run time, fitness value, and error amounts of individual parameters.

T_{60} (s)	0.625	1.25	2.5	5.0
Run time (s)	0.374 (0.046)	0.751 (0.116)	1.607 (0.162)	3.429 (0.372)
Fitness	19.629 (26.107)	9.623 (4.787)	17.997 (6.268)	27.274 (8.432)
ΔT_{60} (s)	0.004 (0.021)	−0.002 (0.022)	−0.029 (0.061)	−0.163 (0.154)
ΔEDT (s)	0.010 (0.010)	0.027 (0.019)	0.083 (0.040)	0.164 (0.080)
ΔC_{80} (dB)	6.232 (2.970)	−0.074 (2.279)	−5.659 (3.191)	−10.974 (3.161)
ΔBR (dB)	−0.015 (2.545)	−0.071 (2.438)	−0.116 (2.514)	0.338 (2.553)

Table 5. Evaluation of the GA using medium quality settings, reporting the mean and standard deviation (in parentheses) of run time, fitness value, and error amounts of individual parameters.

T_{60} (s)	0.625	1.25	2.5	5.0
Run time (s)	0.842 (0.110)	1.692 (0.228)	3.354 (0.347)	6.709 (0.406)
Fitness	13.742 (14.707)	7.712 (4.178)	14.998 (5.940)	23.964 (6.686)
ΔT_{60} (s)	0.002 (0.019)	0.000 (0.016)	−0.010 (0.040)	−0.074 (0.112)
ΔEDT (s)	0.010 (0.010)	0.018 (0.016)	0.058 (0.038)	0.125 (0.078)
ΔC_{80} (dB)	5.359 (3.005)	0.452 (1.980)	−5.148 (3.340)	−10.959 (3.332)
ΔBR (dB)	0.050 (2.170)	0.027 (2.300)	−0.179 (2.455)	0.072 (2.443)

current T_{60} value to determine which array is best suited for the IR, save the new IRs into that specific array, and set that as the filter for the input signal.

Another significant issue is the amount of CPU resources needed to both generate the impulse responses and process the input audio stream in real-time, especially as the length of the IR increases. This is why the native sample rate of the IRs in the plugin are set to 16 kHz for the GA, and why the IRs must then be resampled to the sample rate of the host application to match that of the input audio stream. Regardless, even with a powerful CPU, we were only able to handle IRs with $T_{60} \leq 2.5$ s when running the plugin within Ableton Live 10 at 44.1 kHz before the CPU was overloaded due to the long convolution. It is expected that in the near future the full potential of multi-core processors becomes more accessible for real-time audio plugins so that the partitioned convolution load can be distributed among several cores, preventing in that way the aforementioned overload issues.

Table 6. Evaluation of the GA using high quality settings, reporting the mean and standard deviation (in parentheses) of run time, fitness value, and error amounts of individual parameters.

T_{60} (s)	0.625	1.25	2.5	5.0
Run time (s)	1.514 (0.169)	3.039 (0.268)	6.410 (0.266)	12.499 (0.860)
Fitness	15.009 (15.858)	6.840 (4.413)	13.957 (5.670)	22.125 (6.758)
ΔT_{60} (s)	0.004 (0.019)	0.000 (0.014)	−0.004 (0.043)	−0.069 (0.099)
ΔEDT (s)	0.010 (0.010)	0.016 (0.016)	0.054 (0.036)	0.118 (0.075)
ΔC_{80} (dB)	5.435 (3.028)	0.137 (1.837)	−5.090 (3.163)	−10.734 (3.150)
ΔBR (dB)	−0.193 (2.278)	0.037 (2.153)	−0.117 (2.268)	−0.090 (2.252)

Table 7. Evaluation of the GA using initial settings, reporting the mean (μ) and standard deviation (σ) of run time, fitness value, and error amounts of individual parameters.

	μ	σ
Run time (s)	0.575	0.094
Fitness	1.383	0.750
ΔT_{60} (s)	0.001	0.016
ΔEDT (s)	0.006	0.008
ΔC_{80} (dB)	0.202	0.524
ΔBR (dB)	−0.001	0.106

6 Conclusion

We developed a working prototype for a plugin that aims to create entirely new virtual spaces via a new method that utilizes GAs to create artificial reverberation. The parameters of the plugin have also been carefully chosen and designed so that anyone from music producers to game designers can both understand and utilize the plugin with ease. Evolutionary algorithms can be a source of creativity in digital signal processing, audio design, and music production, and our plugin is one such example. While the IRs generated by the plugin may not always be able to come close to the expected results (within a reasonable amount of time), there is not necessarily a correlation between fitness value and desirability either, and an IR that sounds pleasing to the ear may come at unexpected times. Regardless, further research and development could explore other methods that could decrease computation time and/or fitness value, such as choosing different genetic operations, changing the genetic algorithm parameters, or executing certain tasks in parallel. Adding other methods for generating IRs that model more complex real-life enclosures (without significantly increasing computation time) is also another possibility.

Acknowledgements. The authors thank the Audio Engineering Society for hosting the 2nd ever Matlab Plugin Student Competition at the 147th Convention in New York 2019, Mathworks for providing the travel grants for this competition, and the judges of this competition for their valuable comments.

References

1. Allen, J.B., Berkley, D.A.: Image method for efficiently simulating small-room acoustics. J. Acoust. Soc. Am. **65**(4), 943–950 (1979)
2. Collins, N.: Experiments with a new customisable interactive evolution framework. Organ. Sound **7**(3), 267–273 (2002)
3. Habets, E.: Room impulse response generator. Internal Report, pp. 1–17, January 2006

4. Kristiansen, U., Krokstad, A., Follestad, T.: Extending the image method to higher-order reflections. Appl. Acoust. **38**(2–4), 195–206 (1993)

5. Lochner, J., Burger, J.: The intelligibility of speech under reverberant conditions. Acta Acustica United Acustica **11**(4), 195–200 (1961)

6. Macret, M., Pasquier, P.: Automatic design of sound synthesizers as pure data patches using coevolutionary mixed-typed cartesian genetic programming. In: Proceedings of the 2014 Annual Conference on Genetic and Evolutionary Computation, pp. 309–316. ACM (2014)

7. Mathworks: Matlab. Software (2020). www.mathworks.com. Accessed 20 Feb 2020 (18:18:49)

8. McGovern, S.: Fast image method for impulse response calculations of box-shaped rooms. Appl. Acoust. **70**, 182–189 (2009). https://doi.org/10.1016/j.apacoust.2008.02.003

9. proAV: Room acoustics / concert hall acoustics - an overview. http://www.bnoack.com/acoustic/concerthall.html (1995). Accessed 11 Nov 20 2019

10. Smith, J.O.: Physical Audio Signal Processing. http://ccrma.stanford.edu/jos/pasp/. Accessed 10 Nov 2019. online book, 2010 edition

11. Valimaki, V., Parker, J.D., Savioja, L., Smith, J.O., Abel, J.S.: Fifty years of artificial reverberation. IEEE Trans. Audio Speech Lang. Process. **20**(5), 1421–1448 (2012)

12. Zölzer, U. (ed.): DAFX - Digital Audio Effects, 2nd edn. John Wiley, New York (2011)

Portraits of No One: An Interactive Installation

Tiago Martins[✉] ⓘ, João Correiaⓘ, Sérgio Rebeloⓘ, João Bickerⓘ,
and Penousal Machadoⓘ

CISUC, Department of Informatics Engineering,
University of Coimbra, Coimbra, Portugal
{tiagofm,jncor,srebelo,bicker,machado}@dei.uc.pt

Abstract. Recent developments on artificial intelligence expedited the computational fabrication of visual information, especially photography, with realism and easiness never seen before. In this paper, we present an interactive installation that explores the generation of facial portraits in the borderline between the real and artificial. The presented installation synthesises new human faces by recombining the facial features of its audience and displays them on the walls of the room The array of faces displayed in the installation space is contaminated with real faces to make people question about the veracity of the portraits they are observing. The photo-realism of the generated faces makes it difficult to distinguish the real portraits from the artificial ones.

Keywords: Interactive installation · Media Art · Artificial intelligence · Computer vision · Image generation · Computer graphics

1 Introduction

Portraiture has always been an important form of art. Although its main goal is to immortalise the image of people, it also depicts more than the people traits, thus representing their physical and intellectual possessions [1]. When photography was invented, it soon became the major medium for portraiture by excellence. The portraits, which were formerly an expressive luxury, became affordable for almost everyone [2]. Gradually, portraits became a suitable way to prove the identity of someone and, nowadays, they are part of the identification documents present in everyone's wallet. On the other hand, modern artists started to portrait whatever pleased them in the way they like, aligned with the zeitgeist of the time and the aesthetics of the different art movements [3].

Advances in computer vision enabled the development of more sophisticated detection and recognition tools. This enabled the emergence of artificial intelligence techniques that are able to quickly generate imagery, especially photos, with unprecedented realism. Consequently, these techniques allow the generation of fake content, sometimes used for propaganda, influence, and defamatory

ⓒ Springer Nature Switzerland AG 2020
J. Romero et al. (Eds.): EvoMUSART 2020, LNCS 12103, pp. 104–117, 2020.
https://doi.org/10.1007/978-3-030-43859-3_8

Fig. 1. Environment of the installation *Portraits of No One*. Photo © José Paulo Ruas/DGPC 2019.

purposes. In the case of photos, this is a very important issue, since, by tradition, they are considered proofs of evidence [4].

In this paper, we present the interactive Media Art installation *Portraits of No One* (see Fig. 1). This installation generates and displays portraits of human faces by recombining the facial features of its audience. The generated portraits exhibit a high level of photo-realism, living in the borderline between the real and artificial. This leads users to question themselves about the veracity of the faces they are seeing. This installation was designed for *Sonae Media Art Award 2019*, being selected as a finalist artwork and, consequently, exhibited at the National Museum of Contemporary Art, in Lisbon, Portugal.

Portraits of No One employs the generative system presented in the work *X-Faces* [5,6] to create face portraits. Therefore, each portrait is created through the recombination of parts of different faces. These parts are retrieved from the faces of the users and assembled using computer vision and computer graphics techniques. Users are invited to give their faces to the installation using a capturing box that is placed in the middle of the installation room. This capturing box also records a sound sample during each face capturing. This enables the installation to maintain an ever-changing audiovisual environment consisting of ephemeral portraits that are projected on the walls combined with intertwined sounds produced by the real people behind the portraits.

Whenever users give their faces to the installation, they will immediately see parts of their faces contaminating the portraits being displayed on the walls. In a way, one could say that the installation feeds on the faces of people who interact with it, as new faces become feedstock that enables the installation to grow. This enables the development of a symbiotic interaction between users and the installation, wherein the identities of the users become part of the artwork. Also, this creates a more engaging experience for users by allowing them to direct the emergence of new artificial identities in the form of portraits. Relationships and interactions between users are also discovered as the faces of different users are blended together to create portraits of no one.

The remainder of this paper is organised as follows. Section 2 presents related work. Section 3 presents a comprehensive description of the installation, both in terms of hardware and software. Finally, Sect. 4 draws final conclusions and points future work.

2   Related Work

In the Media Arts context, portraiture is widely explored using video capturing, face detection techniques, and custom-made software. Some early examples include, *e.g.*, the installations *Video Narcissus* (1987) by Shaw [7], *Interfaces* (1990) by Kac [8], or *Solitary* (1992) by Biggs [9].

Recently, this subject has become more popular due to the technological advances in computer vision and, consequently, in face detection. In the scope of this work, we are interested in art installations that use these technologies with three main purposes: *(i)* to create more engaging artistic experiences that enable users to feed the artwork; *(ii)* to create interfaces that enable users to interact between themselves; and *(iii)* to generate artificial human faces.

In some interactive installations, face detection methods are employed to gather content to be used in the creation process of the artwork and, at the same time, to create a more engaging artistic experience for the users. A case in point is *Reface*, an interactive installation developed in 2007 by Levin and Lieberman, which generates group portraits of its users [10]. The portraits are generated through the combination of videos' excerpts of mouths, eyes, and brows, from different users. Also, the users interact with the artwork using eye blinks. In 2015, Howorka developed a digital mirror that displays an average face of all people who have been in front of installation [11]. More recently, in 2019, Tawil presented the installation *IDEMixer* that generates live portraits of the participants by blending pairs of faces captured by two different cameras [12]. In the same year, Cecilia Such presented the interactive multimedia performance *I, You, We* [13]. During this performance, users are invited to take a photo. The photographs are, subsequently, rendered and combined between them, and with other visuals, in a live video that runs in the background. Simultaneously, the performer changes the colours of the video through improvisation with a violin and/or a cello.

Some art installations capture the faces of the users to create interfaces that promote novel ways of interaction. An example is the installation *Sharing Faces* developed by McDonald, in 2013 [14]. This installation behaves as a mirror between two locations: Anyang, in South Korea, and Yamaguchi, in Japan. When users approach the installation, in one of the locations, it tracks their facial expressions and matches them, in real-time, with other expressions of other people who have already stood in front of the installation in the other location. The artists Lancel and Maat developed the installation *Saving Face*, in various geographical contexts, between 2012 and 2016 [15]. In this artwork, portraits of users are blended and displayed in public screens. The intensity of each blend is determined by the number of times users made gestures of "take care" on their faces. In 2015, Lozano-Hemmer designed the interactive installation *Level of Confidence* [16], which uses a face recognition system trained to find similarities between a given face and the faces of forty-three disappeared students in a cartel-related kidnapping in Iguala, Mexico, in 2014. Whenever users stand in front of the installation, it captures their faces and unveils the most similar student and how accurate this match is. In a similar way, Daniele developed the installation

This is not Private, which enables users to see some of their facial features in the faces of someone else [17]. When users stand in front of a screen watching an interview, their faces slowly merge with the face of the person who is being interviewed. The intensity of the merge is determined by the similarity of the expressions between the user and the interviewee. This installation can recognise six basic facial expressions in the faces of users.

Furthermore, some artworks are created only with the goal of generating artificial portraits. An example is the *Portraits of Imaginary People* series, developed by Tyka, in 2017 [18]. These artificial portraits are generated with a generative adversarial network trained with thousands of photos of human faces. In 2017, Moura and Ferreira-Lopes developed an installation that generates portraits based on the false-positives results of facial detection processes [19]. This way, this installation evolves images in an unsupervised way from a series of erroneous matches. The outputs resemble ghost faces. Physically, the installation uses two computers placed against each one, wherein one computer generates visual noise and the other tries to find faces. In 2018, Klingemann developed the *Neural Glitch* technique [20]. He developed this technique randomly altering, deleting or exchanging the resulting networks of fully trained generative adversarial networks. He developed a series of artworks using *Neural Glitch*, some of them focused on artificial portraiture generation.

3   *Portraits of No One*

Portraits of No One is an interactive installation that generates and presents artificial photo-realistic portraits. This is achieved by automatically capturing and extracting the facial features of the visitors, followed by their recombination in a visually seamless way.

This installation was exhibited in the *Sonae Media Art Award 2019* held at the National Museum of Contemporary Art, in Lisbon, Portugal. In this exhibition, the installation was enclosed in a room specifically made for the purpose, with six meters long by four meters wide. As depicted in Fig. 2, the space of the installation comprises two main areas: *(i)* input area and *(ii)* output area. The input area, which is located next to the entrance to the room, contains the capturing box attached to one side of a pillar. The capturing box allows users to capture their faces and this way feed the installation. The output area is located at the back of the room, after the input area, and is surrounded by an immersive video projection of twelve meters wide on three walls. This projection contains an array of portraits that are continuously generated by the installation. Each portrait is displayed with a size similar to that of a real face. This area also includes a speaker attached to the ceiling of the room that reproduces an intertwined combination of sounds recorded during the interactions of the users with the capturing box.

The user interaction is simple and is described as follows: *(i)* the user enters the space of the installation; *(ii)* the user approaches the capturing box and observes her/his face on the screen, in a mirror-like fashion; *(iii)* if the face is

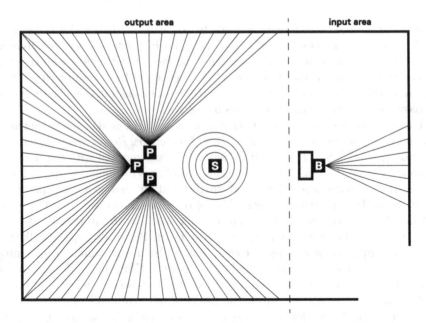

Fig. 2. Floor plan of the installation *Portraits of No One* with the input area (right) and output area (left) along with the main components: capturing box (B), video projectors (P) and speaker (S). Both the video projectors and the speaker are attached to the ceiling of the room. The capturing box is attached to one side of a pillar, with its centre point at a height of 160 cm. There is another key component of the installation, the computer, which is hidden in a space next to the room.

recognised by the system, the button becomes white and blinks fast; otherwise, the button remains red and blinks slowly; *(iv)* when the button is white and the reflected image pleases the user, he/she presses the button to capture the face; *(v)* a countdown of three seconds begins and then the face is captured; *(vi)* the capturing box pauses for a couple of seconds to avoid too frequent captures and the button turns off until a new face can be captured; and, finally, *(vii)* the user steps into the output area to see new portraits being generated using her/his facial parts.

In the following subsections, we describe the body (hardware) and behaviour (software) of the presented installation while considering the corresponding inputs and outputs.

3.1   Hardware

The hardware that builds the installation can be divided into input and output hardware. Input hardware is responsible for collecting data from the user, while output hardware is responsible for displaying audiovisual outputs generated by the installation or providing feedback to the users. The core component in this data flow is the computer, which is responsible for connecting all hardware parts of the installation and managing the inputs and outputs. We use a micro-controller to allow the computer to communicate with the push-button through which users trigger the capture of the face. Figure 3 presents the main hardware components and their data flow.

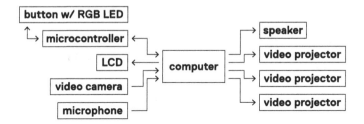

Fig. 3. Diagram of the main hardware components and their data flow. The components on the left are located inside the capturing box, while the ones on the right are attached to the ceiling of the room. The computer is hidden in a space next to the room.

Input Hardware. The input hardware obtains data needed for the installation to operate, including images of users' faces and sounds produced by them. This data is retrieved through the capturing box that contains the following input hardware: *(i)* a video camera to capture the faces; *(ii)* a square LED ring that fits around the camera to provide even illumination with few shadows visible in the captured faces; *(iii)* an LCD display to show users a live preview of their faces before capturing; *(iv)* a microphone to record some seconds of audio during each interaction; and *(v)* a push-button, with an RGB LED, that allows users to trigger the face capturing while providing colour feedback to users.

The video camera and the LED ring are positioned on the top of the display, horizontally centred on the capturing box. The microphone is positioned on one side of the video camera, oriented forward to pick up sounds produced in front of the capturing box. The push-button is placed on the bottom of the screen, also horizontally centred on the capturing box. Figure 4 explains the anatomy of the capturing box.

The capturing box is attached to one side of a pillar, with its centre point at a height of 160 cm. The wall immediately in front of the capturing box is covered with black flannel. This way, we are able to create a completely black background in the portraits. This, in combination with the uniform light produced by the LED ring, provides a coherent photographic style across all portraits. The capturing box was entirely designed and developed in-house, allowing us to control all its features. Figure 5 shows a series of snapshots taken during different stages of the making process, from the laser cutting to the assemblage of custom-made parts that build the capturing box.

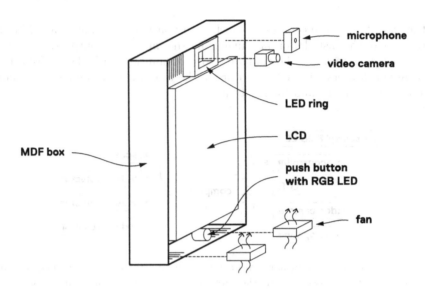

Fig. 4. Schematic of the capturing box showing its anatomy. In addition to the components illustrated in the figure, the box also contains two voltage regulators that allow the adjustment of the brightness of the LED ring and the rotation speed of the fans.

Output Hardware. The output hardware consists of components that allow the installation to express itself to users based on the data it collected from them. This expression is audiovisual and uses images of generated portraits and sounds of different users who gave their faces to the installation. The installation employs three video projectors and one loudspeaker as output hardware to provide an ever-changing audiovisual environment to the users.

The video projectors are attached to the ceiling of the room, each one facing a different wall (see Fig. 2). The video projectors are strategically positioned and oriented to create a single continuous projection. The result is a surrounding video projection of twelve meters wide, on three walls, with a total image size of 5760 by 1080 pixels. This image size allows the presentation of a large array of portraits with proper quality.

The loudspeaker is also attached to the ceiling of the room, in a central position, facing down. This way, we can fill the room with an ambient sound that complements the installation environment.

One could say the LCD screen and the RGB LED on the push-button can also be considered as output hardware, since they provide visual feedback to users while they interact with the capturing box. However, we think that it is easier to make this division between input (capturing box) and output hardware (video projectors and loudspeaker).

Fig. 5. Snapshots of the making process of the capturing box depicting different stages from the laser cutting to the assemblage of custom-made parts, including the body of the box, the adjustable arm for the video camera, and the LED ring that fits around the camera.

3.2 Software

The software that runs the installation is mostly implemented in Java, using the open-source libraries Processing, OpenCV and Dlib. We also use Max for sound recording, processing, and reproduction. The micro-controller that establishes the communication between the computer and the push-button is programmed in Arduino.

The installation relies on a computer system that collects users' data using the input hardware, processes this data, and expresses new portraits through the output hardware.

Input Software. The input software is essentially the part of the system that manages the capturing box (see Fig. 6). One of the key features implemented by this part of the system is real-time face detection. We resort off-the-shelf solutions from OpenCV and Dlib libraries to analyse each frame of the camera feed and detect the faces of users, and their facial landmarks, when they stand in front of the capturing box. In the context of this work, we require neutral frontal poses for the portraits. This is not only advantageous to the accurate extraction of the facial features but also to ensure visual coherence between different portraits. To achieve this, we implemented a set of detection filters that determine whether or not a detected face should be considered as valid.

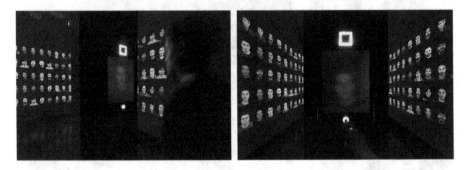

Fig. 6. Photographs of the installation *Portraits of No One* showing someone interacting with the capturing box. On the background of the photographs, one can see walls crowded with faces generated and displayed by the installation system. Photos © José Paulo Ruas/DGPC 2019.

When a valid detection is found, the system draws a rectangle around the face, on the live preview shown on the screen. In situations where more than one valid detection is found, the system focuses on one face based on two preference criteria: *(i)* faces horizontally centred in the captured image; and *(ii)* faces with large bounds.

The rectangle that is drawn around the detected face indicates the area of the captured image that will be cropped and used to generated new portraits. The aspect ratio of the crop rectangle is always 3:4 (vertical). The size and position of the rectangle is determined in relation to the detected facial landmarks. This way, we ensure the faces present in the portraits are always roughly aligned between them.

In addition to drawing a rectangle around the valid face detection, the system invites the user to capture the face by pressing the push-button. To do so, the system changes the colour of the push-button from red, which indicates that no face is being detected, to white and presents a message on the screen encouraging the user to so.

When the user presses the push-button, the system begins a countdown of three seconds, shown on the top of the screen close to the video camera. When the countdown ends, the system captures the face and saves the related data needed to generate new portraits. At this moment, the system turns off the LED of the push-button, during a couple of seconds, until a new face can be captured. This is part of a delay mechanism that we implemented to avoid bursts of photos.

The data that is recorded when a face is captured includes the image of the face, the sound sample recorded from a few seconds before the face being captured to a few seconds later, the points of the facial landmarks that were calculated during detection, and a timestamp indicating the time at which that face was captured. After absorbing and processing the data, the installation can now feed on the new captured face and synthesise new portraits using the generative process described in the next subsection.

Output Software. The output software consists in the part of the system that maintains the audiovisual environment formed by portraits and sounds. This part of the system integrates three modules: the *generator* of artificial portraits based on a set of input faces, the *selector* of faces to generate portraits, and the *producer* of ambient sound.

The *generator* of artificial portraits uses the generative system presented in the work *X-Faces* by Correia et al. in [5,6]. The input of the generator is a set of five faces and their facial landmarks. One face will be our *target* face, *i.e.* the face whose elementary parts (eyebrows, eyes, nose and mouth) will be replaced with parts from other faces. The other four faces will be our *source* faces, *i.e.* the faces that will provide those parts to be blended onto the *target* face. For each face, we calculate the Delaunay triangulation of each face using points based on its landmarks. The triangulation of these points allows us to divide each face image into triangles, which tend to cover corresponding facial features between faces. The next step is to warp each *source* face triangle, *i.e.* the pixels contained in it, to the corresponding *target* face triangle using affine transformations. This results in the *source* faces that are fully aligned with the *target* face. Then, we check if the warped source image part is compatible with the *target*, by calculating the Intersection over Union value of the bounding boxes with a certain threshold. If the value of the Intersection over Union is above the threshold, we consider the part compatible. If they are compatible, we use a seamless cloning algorithm to blend one region of each warped *source* face into the *target* image. The regions of the different facial parts of the warped *source* faces are delimited with masks that are calculated using specific vertexes of the Delaunay triangles. If at least one compatible swap has occurred, the new resulting face is returned. Otherwise, no face is returned. This way, we ensure the seamless swap of facial parts between faces, minimising artefacts and enhancing the realism of the resulting face.

The *selector* of faces manages the faces that are used to compose the grid of 225 portraits that are displayed simultaneously on the walls of the installation. This module selects a random set of faces captured from the users and supplies them to the *generator* to create a new portrait using facial features of these faces. This selection process takes time into account the timestamps of the faces, tending to select faces that are more recent. Each portrait being projected on the walls has a lifespan of 10 to 20 s. After this lifespan is over, the *selector* selects

Fig. 7. Photograph of the installation *Portraits of No One* with someone observing the walls crowded with faces. Photo © José Paulo Ruas/DGPC 2019.

a new set of faces, asks the *generator* to create another artificial portrait, and replaces the old portrait by the new one. This process is repeated whenever the lifespan of a portrait is over. Once in a while, the *selector* chooses a face at random and displays it. This mechanism contaminates the array of faces displayed in the installation with real faces to make users question about the veracity of the portraits they are observing. Also, the smooth transition between portraits makes users aware of the subtle changes on the walls of the installation that are populated with numerous portraits. Figures 7 and 8 show the environment of the installation with users observing the walls populated with the numerous portraits, real and artificial.

The *producer* of ambient sound combines sounds recorded by the capturing box and reproduces the result using the loudspeaker. The sound is generated by intertwining a total of twelve samples randomly selected from the set of all recorded samples. Most of the interactions occur in silence, so we decided to play a base sound sample, in background, at a volume lower than the main samples. This base sound consists in a sample, in loop, of crowd noise that was previously recorded. The frequency and pitch of the recorded sound samples are attenuated and made homogeneous, digitally, through single-pole low-pass sound filters. Generally, the users inside the room mostly hear imperceptible sounds resulting from the mixing of captured noise. However, sometimes, they can recognise some words or even sentences. This provides a sonic environment

Fig. 8. Photograph of the installation *Portraits of No One* with people observing the walls crowded with faces. On the right, one can see the capturing box. Photo © José Paulo Ruas/DGPC 2019.

wherein the user has the perception that the faces depicted in the numerous portraits are talking between them.

4 Conclusions

We have presented the Media Art interactive installation *Portraits of No One*. This installation generates and displays photo-realistic human portraits by recombining facial parts of its audience. This installation was designed for *Sonae Media Art Award 2019* and exhibited at the National Museum of Contemporary Art, in Lisbon, Portugal.

The installation employs the generative system *X-Faces* [5,6] to create the portraits. Each portrait is generated by the recombination of parts of different real faces. These parts are automatically annotated, retrieved, and assembled using techniques from computer vision and computer graphics. The real faces used in this process are portraits of the installation users.

The space of this installation comprises an input and an output area. In the input area, the users are invited to give their faces to the installation, using a capturing box attached to a pillar on the room. In addition to taking pictures, the capturing box also records a sound sample in each interaction. The capturing box was entirely designed, assembled and developed in-house. In the output area, the audiovisual outputs of the installation are presented. Portraits are projected on the walls while ambient sound is reproduced with a loudspeaker. The is generated by the intertwine of sounds of real people interacting with the capturing box. This creates an ever-changing audiovisual environment, contaminated with chaos and doubt.

Whenever users give their faces to the installation, they will immediately see their facial parts contaminating the portraits that are displayed on the walls. The installation feeds on the information of facial portraits from users who interact with it, allowing it to grow. As the installation takes advantage of users to feed itself and grow, the users' facial parts become part of the artwork. This aspect of the installation creates an engaging experience for users by allowing them to directly participate in the generation of new artificial identities in the form of portraits.

Moreover, the artificial portraits generated by the installation encourage critical thinking about how recent technological advances are changing our relationship with the others and the world. The photo-realism of the generated portraits places the audience in the borderline between the real and artificial. As a result, users tend to question the veracity of the faces that they are seeing. Besides, this opens a window of opportunity to discuss the veracity of the images that we see in other contexts and environments and if we should consider them as unquestionable proofs of the truth.

In the installation, one can also observe varied relationships and interactions between users, as well as between the users and the artificial identities generated by the installation. Most of the users envisage the portraits on the walls as other people, starting to judge and talk about their visual appearance. However, when they recognise themselves, or a familiar face part, in a portrait of no one, they change the behaviour. At this moment, users engender empathy by this artificial person and by the others who, like them, are blended to generate their portraits.

Future work will focus on *(i)* set up and test the installation in different locations and *(ii)* further improve the installation system according to feedback already obtained.

Acknowledgments. The authors express their acknowledgement to Directorate General for Cultural Heritage of Portugal, National Museum of Contemporary Art - Museu do Chiado, and Sonae by commissioning the installation *Portraits of No One*. The authors would also like to express their gratitude to the exhibition production staff and to everyone involved in the development of the installation. This work is partially supported by national funds through the Foundation for Science and Technology (FCT), Portugal, within the scope of the project UID/CEC/00326/2019 and it is based upon work from COST Action CA15140: Improving Applicability of Nature-Inspired Optimisation by Joining Theory and Practice (ImAppNIO). The third author is funded by FCT under the grant SFRH/BD/132728/2017.

References

1. Berger, J., Blomberg, S., Fox, C., Dibb, M., Hollis, R.: Ways of Seeing. Penguin Books Ltd., London (1972)

 2. Benjamin, W.: A short history of photography. Screen **13**(1), 5–26 (1972)
 3. West, S.: Portraiture and modernism. In: Oxford History of Art: Portraiture, chap. 8, pp. 187–204. Oxford University Press, Oxford (2004)
 4. Sontag, S.: In Plato's cave. In: On Photography, chap. 1, pp. 1–19. Farrar, Straus & Giroux, New York (1973)
 5. Correia, J., Martins, T., Martins, P., Machado, P.: X-faces: the eXploit is out there. In: Pachet, F., Cardoso, A., Corruble, V., Ghedini, F. (eds.) Proceedings of the Seventh International Conference on Computational Creativity (ICCC 2016), pp. 164–182. Sony CSL, Paris (2016)
 6. Correia, J., Martins, T., Machado, P.: Evolutionary data augmentation in deep face detection. In: GECCO 2019 - Proceedings of the 2019 Genetic and Evolutionary Computation Conference, Prague, Czech Republic (2019)
 7. Shaw, J.: Video narcissus (1987). https://www.jeffreyshawcompendium.com/portfolio/video-narcissus/. Accessed 9 Nov 2019
 8. Kac, E.: Interfaces (1990). http://www.ekac.org/sstv.html. Accessed 10 Nov 2019
 9. Biggs, S.: Solitary (1992). http://littlepig.org.uk/installations/solitary/solitary.htm. Accessed 10 Nov 2019
10. Levin, G., Lieberman, Z.: Reface [portrait sequencer] (2007). http://www.flong.com/projects/reface/. Accessed 8 Nov 2019
11. Howorka, S.: Average face mirror (2015). http://www.sarahhoworka.at/projects/average-face-mirror. Accessed 10 Nov 2019
12. Tawil, S.A.: Idemixer (2019). http://samehaltawil.com/portfolio/idemixer/. Accessed 10 Nov 2019
13. Suhr, H.C.: I, you, we: exploring interactive multimedia performance. In: Proceedings of the 27th ACM International Conference on Multimedia, MM 2019, pp. 1147–1148. ACM, New York (2019)
14. McDonald, K.: Sharing faces (2013). https://github.com/kylemcdonald/sharingfaces. Accessed 10 Nov 2019
15. Lancel, K., Maat, H., Brazier, F.: Saving face: playful design for social engagement, in public smart city spaces. In: Brooks, A.L., Brooks, E., Sylla, C. (eds.) ArtsIT 2018, DLI 2018. LNICST, vol. 265, pp. 296–305. Springer, Cham (2019). https://doi.org/10.1007/978-3-030-06134-0_34
16. Lozano-Hemmer, R.: Level of confidence (2015). http://www.lozano-hemmer.com/artworks/level_of_confidence.php. Accessed 10 Nov 2019
17. Daniele, A.: This is not private (2015). http://www.letitbrain.it/letitbrain/?port=this-is-not-private. Accessed 10 Nov 2019
18. Tyka, M.: Portraits of imaginary people (2017). http://www.miketyka.com/?s=faces. Accessed 10 Nov 2019
19. Moura, J.M., Ferreira-Lopes, P.: Generative face from random data, on 'how computers imagine humans'. In: Proceedings of the 8th International Conference on Digital Arts, ARTECH 2017, pp. 85–91. ACM, New York (2017). https://doi.org/10.1145/3106548.3106605
20. Klingemann, M.: Neural glitch (2018). http://underdestruction.com/2018/10/28/neural-glitch/. Accessed 10 Nov 2019

Understanding Aesthetic Evaluation
Using Deep Learning

Jon McCormack[1]([✉]) [iD] and Andy Lomas[2] [iD]

[1] SensiLab, Monash University, Caulfield East 3145, VIC, Australia
Jon.McCormack@monash.edu
[2] Goldsmiths, University of London, London SE14 6NW, UK
a.lomas@gold.ac.uk
http://sensilab.monash.edu
https://www.gold.ac.uk/computing

Abstract. A bottleneck in any evolutionary art system is aesthetic evaluation. Many different methods have been proposed to automate the evaluation of aesthetics, including measures of symmetry, coherence, complexity, contrast and grouping. The interactive genetic algorithm (IGA) relies on human-in-the-loop, subjective evaluation of aesthetics, but limits possibilities for large search due to user fatigue and small population sizes. In this paper we look at how recent advances in deep learning can assist in automating personal aesthetic judgement. Using a leading artist's computer art dataset, we use dimensionality reduction methods to visualise both genotype and phenotype space in order to support the exploration of new territory in any generative system. Convolutional Neural Networks trained on the user's prior aesthetic evaluations are used to suggest new possibilities similar or between known high quality genotype-phenotype mappings.

Keywords: Evolutionary art · Aesthetics · Aesthetic measure ·
Convolutional Neural Networks · Dimension reduction · Morphogenesis

1 Introduction

Artistic evolutionary search systems, such as the Interactive Genetic Algorithm (IGA) have been used by artists and researchers for decades [3,4,8,23,25,27, 29,31,32]. A key advantage of the IGA is that it substitutes formalised fitness measures for human judgement. The algorithm arose to circumvent the difficulty in developing generalised fitness measures for "subjective" criteria, such as personal aesthetics or taste. Hence the IGA found favour from many artists and designers, keen to exploit the powerful search and discovery capabilities offered by evolutionary algorithms, but unable to formalise their aesthetic judgement in computable form.

Over the years, the research community has proposed many new theories and measures of aesthetics, with research from both the computational aesthetics (CA) and psychology communities [13]. Despite much effort and many

© Springer Nature Switzerland AG 2020
J. Romero et al. (Eds.): EvoMUSART 2020, LNCS 12103, pp. 118–133, 2020.
https://doi.org/10.1007/978-3-030-43859-3_9

advances, a computable, universal aesthetic measure remains elusive – an open problem in evolutionary music and art [24]. One of the reasons for this is the psychological nature of aesthetic judgement and experience. In psychology, a detailed model of aesthetic appreciation and judgement has been developed by Leder and colleagues [14,15]. This model describes information-processing relationships between various components that integrate into an aesthetic experience and lead to an aesthetic judgement and aesthetic emotion. The model includes perceptual aesthetic properties, such as symmetry, complexity, contrast, and grouping, but also social, cognitive, contextual and emotional components that all contribute in forming an aesthetic judgement. A key element of the revised model [15] is that it recognises the influence of a person's affective state on many components and that aesthetic judgement and aesthetic emotion co-direct each other.

One of the consequences of this model is that any full computational aesthetic measure must take into account the affective state of the observer/participant, in addition to other factors such as previous experience, viewing context and deliberate (as opposed to automatic) formulations regarding cognitive mastering, evaluation and social discourse. All factors that are very difficult or impossible for current computational models to adequately accommodate.

How then can we progress human-computer collaboration that involves making aesthetic judgements if fully developing a machine-implementable model remains illusive? One possible answer lies in teaching the machine both tacit and learnt knowledge about an individual's personal aesthetic preferences so that the machine can assist a person in creative discovery. The machine provides assistance only, it does not assume total responsibility for aesthetic evaluation.

In this paper we investigate the use of a number of popular machine learning methods to assist digital artists in searching the large parameter spaces of modern generative art systems. The aim is for the computer to learn about an individual artist's aesthetic preferences and to use that knowledge to assist them in finding more appropriate phenotypes. "Appropriate" in the sense that they fit the artist's conception of high aesthetic value, or that they are in some category that is significant to the artist's creative exploration of a design space. Additionally, we explore methods to assist artists in understanding complex search spaces and use that information to explore "new and undiscovered territory". Finally, we discuss ways that mapping and learning about both genotype and phenotype space can inspire a search for new phenotypes "in between" known examples. These approaches aim to eliminate the user fatigue of traditional IGA approaches.

1.1 Related Work

In recent years, a variety of deep learning methods have been integrated into evolutionary art systems. Blair [5] used adversarial co-evolution, evolving images using a GP-like system alongside a LeNet-style Neural Network critic.

Bontrager et al. [6] describe an evolutionary system that uses a Generative Adversarial Network (GAN), putting the latent input vector to a trained GAN under evolutionary control, allowing the evolution of high quality 2D images in a target domain.

Singh et al. [30] used the feature vector classifier from a Convolutional Neural Network (CNN) to perform rapid visual similarity search. Their application was for design inspiration by rapidly searching for images with visually similar features from a target image, acquired via a smartphone camera. The basis of our method is similar in the use of using a large, pre-trained network classifier (such as ResNet-50) to find visual similarity between generated phenotype images and a database of examples, however our classifier is re-trained on artist-specific datasets, increasing its accuracy in automating personal aesthetic judgement.

2   Exploring Space in Generative Design

In the experiments described in this paper, we worked with a dataset of evolutionary art created by award-winning artist Andy Lomas. Lomas works with developmental morphogenetic models that grow and develop via a cellular development process. Details of the technical mechanisms of his system can be found in [18]. A vector of just 12 real valued parameters determines the resultant form, which grows from a single cell into complex forms often involving more than one million cells. The range of forms is quite varied, Fig. 1 shows a small selection of samples. In exploring the idea of machine learning of personal aesthetics, we wanted to work with a real, successful artistic system[1], rather than an invented one, as this allows us to understand the ecological validity [7] of any system or technique developed. Ecological validity requires the assessment of creative systems in the typical environments and contexts under which they are actually experienced, as opposed to a laboratory or artificially constructed setting. It is considered an important methodology for validating research in the creative and performing arts [12].

Fig. 1. Example cellular forms generated by cellular morphogenesis

---

[1] Lomas is an award winning computer artist who exhibits internationally, see his website http://www.andylomas.com.

2.1   Generative Art Dataset

The dataset used consisted of 1,774 images, each generated by the developmental form generation system using the software *Species Explorer* [19]. Each image is a two-dimensional rendering of a three-dimensional form that has been algorithmically grown based on 12 numeric parameters (the "genotype"). The applied genotype determines the final developed 3D form (the "phenotype"), which is rendered by the system to create a 2D image of the 3D form.

The dataset also contains a numeric aesthetic ranking of each form (ranging from 0 to 10, with 1 the lowest and 10 the highest, 0 meaning a failure case where the generative system terminated without generating a form). These rankings were all performed by Lomas, so represent his personal aesthetic preferences. Ranking visual form in this manner is an integral part of using his Species Explorer software, with the values in the dataset created over several weeks as he iteratively generated small populations of forms, ranked them, then used those rankings to influence the generation of the next set of forms.

Lomas has also developed a series of stylistic categorisations that loosely describe the visual class that each form fits into. This categorisation becomes useful for finding forms *between* categories, discussed later. Category labels included "brain" (245 images), "mess" (466 images), "balloon" (138 images), "animal" (52 images), "worms" (32 images) and "no growth" (146 images). As each form develops through a computational growth simulation process, some of the images fail to generate much at all, leading to images that are "empty" (all black or all white). There were 252 empty images, leaving 1,522 images of actual forms. Even though the empty images are not visually interesting, they still hold interesting data as their generation parameters result in non-viable forms. Most of the category data (1,423 images) had been created at the same time as when Lomas was working on the original Cellular Forms. The remaining 351 images were categorised by Lomas as part of this project.

2.2   Understanding the Design Space

As a first step in understanding the design space we used a variety of dimension reduction algorithms to visualise the distribution of both genotype and phenotype space to see if there was any visible clustering related to either aesthetic ranking scores or categories. We experimented with a number of different algorithms, including t-SNE [20], UMAP [26] and Variational Autoencoders [21], to see if such dimension reduction visualisation techniques could help artists better understand relationships between genotype and categories or highly ranked species.

As shown in Fig. 2, the dimensionally reduced genotype space tends to have little visible structure. The figure shows each 12-dimensional genotype dimensionally reduced to two dimensions and colour-coded according to category (top) and rating (bottom). In the case of rating, we reduced the ten-point numeric scale to five bands for clarity. The figure shows the results obtained with the

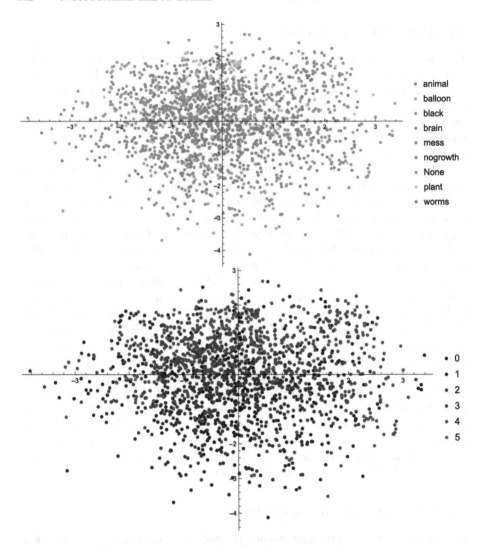

Fig. 2. Plot of genotype distribution in 2 dimensions using t-SNE. Individual genotypes are coloured by category (top) and by score (bottom).

t-SNE dimension reduction, testing with other algorithms (PCA, UMAP and a Variational Autoencoder) did not result in significantly better visual results.

Although some grouping can be seen in the figure, any obvious overall clustering is difficult to observe, particularly for the categories. While there is some overall structure in the score visualisation (high ranked individuals tend to concentrated in the around the upper left quadrant), discerning any regions of high or low quality is difficult. In many cases, low and high ranked individuals map to close proximity in the 2D representation.

What this analysis reveals is that the genotype space is highly unstructured in relation to aesthetic concerns, making it difficult to easily evolve high quality phenotypes. The developmental nature of the generative system, which depends on physical simulation, means that small parameter changes at critical points can result in large differences in the resultant developed form.

2.3   Phenotype Space

To visualise the phenotype space we used the feature classification layer of the ResNet-50 convolutional neural network. Because ResNet was trained on 1.2 million images from the ImageNet dataset [9], it is very good at identifying image features that humans also recognise. Networks trained on the ImageNet classification tasks have been shown to work very well as off the shelf image extractors [28], and show even better results when fine-tuned to datasets for the task at hand [1]. The network produces a 2048-element vector based on the features of an input image. This vector is then dimensionally reduced to create a two-dimensional visualisation of the feature space. Again, we used the t-SNE algorithm to reduce the dimensionality of the space.

Figure 3 shows the results for both the category (top) and score (bottom) classifications. As the figure shows, this time structure can be seen in the feature data. Classifications such as "black" and "balloon" are visible in specific regions. Similarly, the score distribution shows increasing values towards the upper-right quadrant in the visualisation.

Such visualisations can therefore potentially assist artists in navigating and understanding the space of possibilities of their generative system, because they allow them to direct search in specific regions of phenotype space. A caveat here is that the dimension reduction process ideally needs to be reversible, i.e. that one can go from low dimensions to higher if selection specific regions on a 2D plot. As a minimum, it is possible to determine a cluster of nearby phenotypes in 2D space and seed the search with the genotypes that created them, employing methods such as hill climbing to search for phenotypes with similar features.

2.4   Parameter Searching and Interpolation

In early work, such as Lomas' Aggregation [17] and Flow [16] series, the artist would create plots showing how the phenotype changes depending on parameter values of the genotype. An example of such a plot can be seen in Fig. 4. In these systems the genotype had a very low number of dimensions, typically just two or three parameters, which allowed a dense sampling of the space of possibilities by simply independently varying each parameter in the genotype over a specified range, running the generative system with each set of parameter values, and plotting the results in a chart with positions for each image based on the genotype parameters. One intuition from these plots is that the most interesting rich and complex behaviour often happens at transition points in the genotype space, where one type of characteristic behaviour changes into another. This can be seen in Fig. 4 where the forms in the 6th and 7th columns

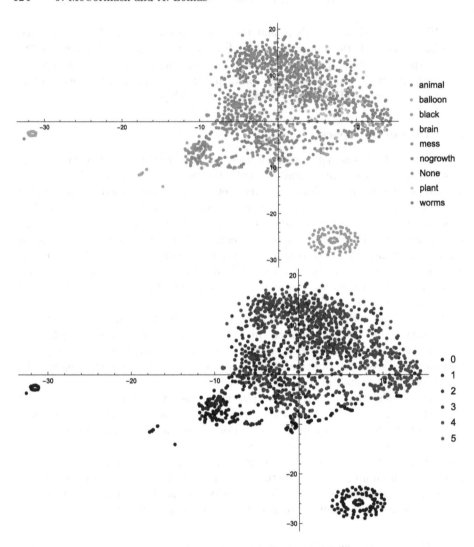

Fig. 3. Plot of phenotype distribution in 2 dimensions using t-SNE. Individual phenotypes are coloured by category (top) and by score (bottom).

are particularly richly structured. These changes occur at parameter settings where the generative system was at a transition state between stability (to the left) and instability (to the right).

As the number of dimensions increases performing a dense sampling of the genotype space runs into the "Curse of Dimensionality" [2,10], where the number of samples needed increases exponentially with the number of parameters. Even if enough samples can be taken, how to visualise and understand the space becomes difficult and concepts such as finding the nearest neighbours to any point in the parameter space become increasingly meaningless [22]. One potential

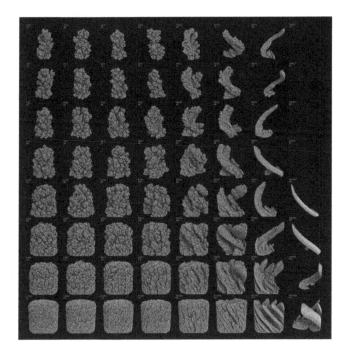

Fig. 4. Plot from Aggregation series showing effects of varying genotype parameters

approach to make sense of higher dimensional spaces is to categorise different phenotypes created by the system. By defining categories for phenotypes we can express searching for transition points in a meaningful way as being the places in genotype space where small changes in the genotype result in changing from one phenotype category to another.

3   Learning an Artist's Aesthetic Preferences

A ResNet-50 classifier was tested with the same dataset of 1,774 images with ratings and categories as described above, using 1,421 images in the training dataset and 353 images in the validation dataset. Re-training the final classifier layers of ResNet-50 created a network that matched Lomas' original categories in the validation set with an accuracy of 87.0%. We also looked at the confidence levels for the predictions, based on the difference between the network's probability value for the predicted category and the probability level of the highest alternative category.

Table 1 shows how the prediction accuracy varies depending on the confidence levels. The network has a reliability of over 97% for the images in the top two confidence quartiles, with 69% of the incorrect categorisations being in the lowest confidence quartile. A visual inspection of images in the lowest confidence quartile confirmed that these were typically also less clear which category an image should be put in to a human observer.

Table 1. ResNet-50 accuracy levels for different confidence quartiles.

Confidence quartile	Prediction accuracy
75% to 100%	97.1%
50% to 75%	97.9%
25% to 50%	90.5%
0% to 25%	67.6%

The confusion matrix in Fig. 5 shows that the predictions appear to be consistently good across all categories, with the majority of items in each category predicted correctly. The most confused categories were "mess" and "nogrowth", both of which indicate forms that are considered by the artist to be aesthetic failure cases and sometimes look quite similar.

As well as being used for categorisation, a ResNet-50 network with a scalar output was trained against the aesthetic ranking using values from 0 to 10 that Lomas had given the forms. This resulted in a network that predicted the ranking of images in the validation set with a root mean square error of 0.716. Given that these ranking are subjective evaluations of images that often have very similar appearance this appears to be a high level of predictive accuracy.

3.1 Genotype Space

The dataset was tested to see whether predictions of the phenotype category and aesthetic rank could be obtained from genotype parameters. This is desirable as good predictions of phenotype from the genotype values could directly aid exploration of the space of possibilities. Techniques such as Monte Carlo methods could be used to choose new candidate points in genotype space with specified fitness criteria. We could use the predictions to generate plots of expected behaviour as genotype parameters are varied that could help visualise the phenotype landscape and indicate places in genotype space where transitions between phenotype classes may occur. If meaningful gradients can be calculated from predictions, gradient descent could be used to directly navigate towards places in genotype space were one category is predicted to change into another and transitional forms between categories may exist.

Fast.ai [11], a Python machine learning library to create deep learning neural networks, was used to create neural net predictors for the category and aesthetic rank using genotype values as the input variables. The fast.ai Tabular model was used, with a configuration of two fully connected hidden layers of size 200 and 100. The same training and validation sets were used as previously.

Using these neural nets we achieved an accuracy of 68.3% for predictions of the category, and predictions of the aesthetic rank had a root mean square error of 1.88. These are lower quality predictions that we obtained with the ResNet-50 classifier using the phenotype, but this is to be expected given that the ranking and categorisation are intended to be evaluations of the phenotype and were

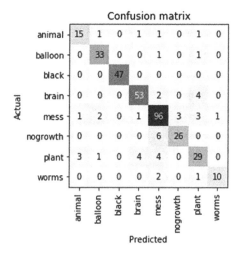

Fig. 5. Confusion matrix for ResNet-50 (phenotype space) categorisor

done by Lomas looking at the images of the phenotype forms. The results are also confirmed by the dimensionally-reduced visualisations presented in Sect. 2.2.

The confusion matrix for the category predictions is shown in Fig. 6. Similarly to the results with the ResNet-50 (phenotype space) categoriser, the "mess" and "nogrowth" categories are often confused, but with the genotype space categoriser the "plant" and "brain" categories are also quite frequently confused with each other. This suggests that it might be worth generating more training data for the "brain" and "plant" categories to improve predictive accuracy, but could also be an indication that the "plant" and "brain" categories are closely connected in genotype space.

The current version of Lomas' "Species Explorer" software uses a simple k-Nearest Neighbours (k-NN) method to give predictions of phenotype based on genotype data [18]. Testing with the same validation set, the k-NN method predicts categories with an accuracy of 49.8%, and the aesthetic rank with a mean square error of 2.78. The genotype space neural net predictors give significantly better predictions than the k-NN predictor.

A new feature was added into Species Explorer that uses the genotype neural network predictors to generate 2D cross-section plots through genotype space, showing the predicted categories or rank at different coordinate positions, see Fig. 7. As can be seen, these plots predict a number of potential places where transitions between categories may occur, which could lead the artist to explore new regions of the genotype space.

4   Discussion

Our results of incorporating this new search feature into an artist's creative workflow indicate that deep learning based neural nets appear to be able to achieve

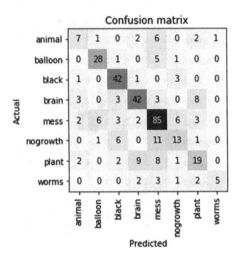

Fig. 6. Confusion matrix for Tabular (genotype space) categoriser

good levels of accuracy when predicting the phenotype categories and aesthetic rank evaluations made by Lomas in the test dataset. The best predictions were achieved with ResNet-50, a pre-trained convolutional neural network designed for image recognition, using phenotype image data as the input. Additionally, we achieved potentially useful levels of prediction from genotype data using the fast.ai library's Tabular model to create a deep learning neural net with two fully connected hidden layers. Predictions based on the genotype rather than the phenotype are particularly interesting as they should allow navigation directly in genotype space to suggest new points to sample.

One of the main reasons for using IGAs is that the fitness function is unknown, or may not even be well defined because the artist's judgement changes over time. The use of the neural networks in this work can be seen as trying to discover whether there is a function that matches the artist's aesthetic evaluations with a useful level of predictability. If such a function can be found it could be used in a number of ways, such to use monte carlo sampling along with providing a fitness function for conventional evolutionary algorithms. If the discovered fitness function is sufficiently simple (such as being unimodal) methods like hill climbing may be appropriate.

Lomas has been using a k-Nearest Network in Species Explorer to give prediction based on position in genotype space. As the numbers of dimensions increase k-NN performance generally becomes significantly less effective [22], while deep neural networks can still be effective predictors with higher dimensional inputs. This means the deep neural networks have the potential to allow useful levels of prediction from genotype space in systems with high numbers of genotype parameters.

It is likely that with more training data we will be able to improve the predictions from genotype space. This raises the possibility for a hybrid system:

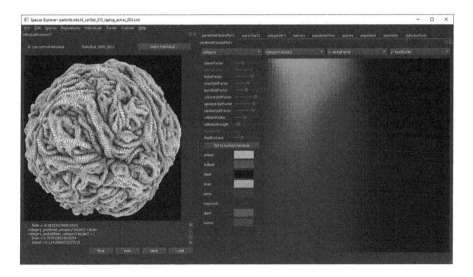

Fig. 7. Species Explorer user interface, showing 2D cross-section plots through geno-type space using a neural net to predict the phenotype category at new positions in genotype space

if we have a convolutional neural network that can achieve high levels of accuracy from phenotype data we could use this to automate creation of new training data for a genotype space predictor. In this way, improving the ability of a genotype space based predictor may be at least partially automated.

There is a lot of scope for trying out different configurations of deep neural networks for genotype space predictions. The choice of a network in this study, with two fully connected hidden layers with 200 and 100 neurons, was simply based on the default values suggested in the fast.ai documentation of their Tab-ular model. A hyper-parameter search would likely reveal even better results.

An important part of this process is how to make ranking and categorisa-tion as easy for a creative practitioner as possible. The aim should be to allow the artist to suggest new categories and ways of ranking with as few training examples as are necessary to get good levels of prediction. It should also facili-tate experimentation, making the process of trying out new ways of ranking and different ways of categorising behaviour as simple as possible.

In both authors' experience, it is often only after working with a generative system for some time, typically creating hundreds of samples, that categories of phenotype behaviour start to become apparent. This means that manually categorising all the samples that have already been generated can become sig-nificantly laborious. This has meant that although Lomas' Species Explorer soft-ware allows phenotype samples to be put into arbitrary categories, and data from categorisation can be used to change fitness functions used to generate new sam-ples, for the majority of systems Lomas has created he hasn't divided phenotype results into categories and has relied on aesthetic rank scores instead. This is

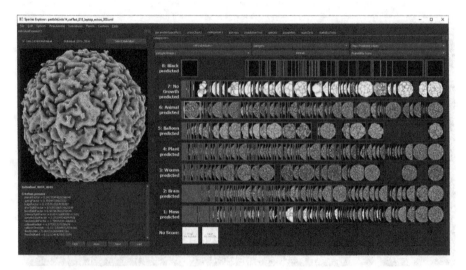

Fig. 8. Species Explorer user interface, showing predicted categorisations from a ResNet-50 network. The items are ordered based on the confidence levels for the predicted category, with the highest confidence level predictions to the left of each group

one area where pre-trained network classifiers, such as ResNet-50, may be useful. If we can reliably train a neural network to classify different phenotypes with only a small amount of training data it could make the process of creating and testing different ways of categorising phenotypes significantly easier.

We modified the existing user interface in Species Explorer so that predictions of how a classifier would divide data into classes can be shown, together with placement and colouring of the outlines of thumbnails based on the confidence levels of predictions, see Fig. 8. This allows a simple evaluation of the quality of prediction, and helps indicate samples that might be good to add to the training set (such as incorrect predictions that the classifier has done with high confidence) to improve the quality of predictions.

The tests with dimensionally reduced plots in phenotype space using t-SNE on the feature vectors of ResNet-50 appear to show meaningful structure which may be useful to help divide samples into categories. In particular, this technique may be useful both to help initial categorisation, broadly dividing samples in phenotype space into categories, and to help sub-divide existing categories that the user wants to explore separating into different classes. The use of plots such as these may actively help experimentation, allowing the creative users to modify existing classification schemes and quickly try out different ideas of how to categorise phenotypes.

5   Conclusions and Future Work

The aim of this research was to progress machine-assisted aesthetic judgement based on an artist's personal aesthetic preferences. We worked with an established artist's work to give ecological validity to our system. While the results are specific to an individual artist, it is worth emphasising that the methods discussed generalise to any multi-parameter generative system whose phenotypes can be expressed as 2D images. Indeed, the Species Explorer software separates the creative evolution process from the actual generative system, allowing Species Explorer to work with *any* parameter based generative system.

The research presented here shows that deep learning neural networks can be useful to predict aesthetically driven evaluations and assist artists to find phenotypes of personally high aesthetic value. As discussed, these predictors are useful to help explore the outputs of generative systems directly in genotype space.

There is still more research to be done however. More testing is now needed to see how productive this is in practice when working with systems that often have high dimensional parameter spaces. We have shown the neural networks can categorise and rank phenotypes with a high accuracy in a specific instance, the next step would be to see if this approach generalises to other artists and their personal aesthetics.

Acknowledgements. This research was supported by an Australian Research Council grant FT170100033 and a Monash University International Research Visitors Collaborative Seed Fund grant.

References

1. Azizpour, H., Sharif Razavian, A., Sullivan, J., Maki, A., Carlsson, S.: From generic to specific deep representations for visual recognition. In: Proceedings of the IEEE Conference on Computer Vision and Pattern Recognition Workshops, pp. 36–45 (2015)
2. Bellman, R.E.: Adaptive Control Processes: A Guided Tour. Princeton University Press, Princeton (1961)
3. Bentley, P.J.: Evolutionary Design by Computers. Morgan Kaufmann Publishers, San Francisco (1999)
4. Bentley, P.J., Corne, D.W. (eds.): Creative Evolutionary Systems. Academic Press, London (2002)
5. Blair, A.: Adversarial evolution and deep learning – how does an artist play with our visual system? In: Ekárt, A., Liapis, A., Castro Pena, M.L. (eds.) EvoMUSART 2019. LNCS, vol. 11453, pp. 18–34. Springer, Cham (2019). https://doi.org/10. 1007/978-3-030-16667-0_2
6. Bontrager, P., Lin, W., Togelius, J., Risi, S.: Deep interactive evolution. In: Liapis, A., Romero Cardalda, J.J., Ekárt, A. (eds.) EvoMUSART 2018. LNCS, vol. 10783, pp. 267–282. Springer, Cham (2018). https://doi.org/10.1007/978-3-319-77583-8_18
7. Brunswik, E.: Perception and the Representative Design of Psychological Experiments, 2nd edn. University of California Press, Berkley and Los Angeles (1956)
8. Dawkins, R.: The Blind Watchmaker. No. 332, Longman Scientific & Technical, Essex (1986)

9. Deng, J., Dong, W., Socher, R., Li, L.J., Li, K., Fei-Fei, L.: ImageNet: a large-scale hierarchical image database. In: 2009 IEEE Conference on Computer Vision and Pattern Recognition, CVPR 2009, pp. 248–255. IEEE (2009)
10. Donoho, D.L., et al.: High-dimensional data analysis: the curses and blessings of dimensionality. AMS Math Chall. Lect. **1**(2000), 32 (2000)
11. Howard, J., et al.: Fastai (2018). https://github.com/fastai/fastai
12. Jausovec, N., Jausovec, K.: Brain, creativity and education. Open Educ. J. **4**, 50–57 (2011)
13. Johnson, C.G., McCormack, J., Santos, I., Romero, J.: Understanding aesthetics and fitness measures in evolutionary art systems. Complexity **2019**, 14 (2019). https://doi.org/10.1155/2019/3495962. Article ID 3495962
14. Leder, H., Belke, B., Oeberst, A., Augustin, D.: A model of aesthetic appreciation and aesthetic judgments. Br. J. Psychol. **95**, 489–508 (2004)
15. Leder, H., Nadal, M.: Ten years of a model of aesthetic appreciation and aesthetic judgments: the aesthetic episode - developments and challenges in empirical aesthetics. Br. J. Psychol. **105**, 443–464 (2014)
16. Lomas, A.: Flow. http://www.andylomas.com/flow.html
17. Lomas, A.: Aggregation: complexity out of simplicity. In: ACM SIGGRAPH 2005 Sketches, p. 98. ACM (2005)
18. Lomas, A.: Cellular forms: an artistic exploration of morphogenesis. In: AISB 2014–50th Annual Convention of the AISB, July 2014
19. Lomas, A.: Species explorer: an interface for artistic exploration of multi-dimensional parameter spaces. In: Bowen, J., Lambert, N., Diprose, G. (eds.) Electronic Visualisation and the Arts (EVA 2016). Electronic Workshops in Computing (eWiC), 12th–14th July 2016. BCS Learning and Development Ltd., London (2016)
20. van der Maaten, L., Hinton, G.: Visualizing data using t-SNE. J. Mach. Learn. Res. **9**(Nov), 2579–2605 (2008)
21. Makhzani, A., Shlens, J., Jaitly, N., Goodfellow, I., Frey, B.: Adversarial autoencoders. arXiv preprint arXiv:1511.05644 (2015)
22. Marimont, R., Shapiro, M.: Nearest neighbour searches and the curse of dimensionality. IMA J. Appl. Math. **24**(1), 59–70 (1979)
23. McCormack, J.: Interactive evolution of forms. In: Cavallaro, A., Harley, R., Wallace, L., Wark, M. (eds.) Cultural Diversity in the Global Village: Third International Symposium on Electronic Art, p. 122. The Australian Network for Art and Technology, Sydney (1992)
24. McCormack, J.: Open problems in evolutionary music and art. In: Rothlauf, F., et al. (eds.) EvoWorkshops 2005. LNCS, vol. 3449, pp. 428–436. Springer, Heidelberg (2005). https://doi.org/10.1007/978-3-540-32003-6_43
25. McCormack, J.: Creative systems: a biological perspective. In: Veale, T., Cardoso, F. (eds.) Computational Creativity. CSCS, pp. 327–352. Springer, Cham (2019). https://doi.org/10.1007/978-3-319-43610-4_15
26. McInnes, L., Healy, J., Melville, J.: UMAP: uniform manifold approximation and projection for dimension reduction. arXiv preprint arXiv:1802.03426 (2018)
27. Rowbottom, A.: Evolutionary art and form. In: Evolutionary Design by Computers, pp. 261–277. Morgan Kaufmann, San Francisco (1999)
28. Sharif Razavian, A., Azizpour, H., Sullivan, J., Carlsson, S.: CNN features off-the-shelf: an astounding baseline for recognition. In: Proceedings of the IEEE Conference on Computer Vision and Pattern Recognition Workshops, pp. 806–813 (2014)
29. Sims, K.: Artificial evolution for computer graphics. In: Computer Graphics, vol. 25, pp. 319–328. ACM SIGGRAPH, New York, July 1991. http://www.genarts.com/karl/papers/siggraph91.html

30. Singh, D., Rajcic, N., Colton, S., McCormack, J.: Camera obscurer: generative art for design inspiration. In: Ekárt, A., Liapis, A., Castro Pena, M.L. (eds.) Evo-MUSART 2019. LNCS, vol. 11453, pp. 51–68. Springer, Cham (2019). https://doi.org/10.1007/978-3-030-16667-0_4
31. Todd, S., Latham, W.: Mutator: a subjective human interface for evolution of computer sculptures. Technical report (1991)
32. Todd, S., Latham, W.: Evolutionary Art and Computers. Academic Press, London (1992)

An Aesthetic-Based Fitness Measure and a Framework for Guidance of Evolutionary Design in Architecture

Manuel Muehlbauer[1(✉)] [iD], Jane Burry[2], and Andy Song[3]

[1] FutureImmersion, Lohr a. Main, Germany
info@futureimmersion.com
[2] Swinburne University, Hawthorn 3122, Australia
jburry@swin.edu.au
[3] Evolutionary Computation and Machine Learning Group,
RMIT University Melbourne, Carlton 3000, Australia
andy.song@rmit.edu.au
https://titan.csit.rmit.edu.au/~e46507/ecml/

Abstract. The authors present an interactive design framework for grammar evolution. A novel aesthetic-based fitness measure was introduced as guidance procedure. One feature of this guidance procedure is the initial input of a reference image. This image provides direction to the evolutionary design process by application of a similarity measure. A case study shows the interactive exploration of a set of 3D-shapes using the presented framework for guidance of evolutionary design in architecture. The aesthetic-based fitness measure combing quantitative and qualitative criteria was applied in this case study to evaluate form and function of architectural design solutions.

Keywords: Evolutionary computation · Architectural design · Grammar evolution · Fitness evaluation

1 Introduction and Literature Review

The presented research aimed at exploring a certain design space for architectural shapes in a more efficient way than other evolutionary design systems. Features of the architectural design process were explored, looking for new ways of integrating the user in the evolutionary cycle. An aesthetic-based fitness measure was investigated to guide the evolutionary process. Two main features for interactive guidance were explored and tested in a case study. These features called Shape Comparison and Online Classification were investigated based on a similarity measure. This similarity measure compared the feature vectors of the initial image and design solutions. An adaptive system based on genetic programming was used to account for the divergent character of architectural design exploration.

In the literature regarding adaptive systems design exploration was described as providing a user with a way to experience a variety of design solutions [1]. Several algorithms for shape generation were analysed regarding their potential and implications

© Springer Nature Switzerland AG 2020
J. Romero et al. (Eds.): EvoMUSART 2020, LNCS 12103, pp. 134–149, 2020.
https://doi.org/10.1007/978-3-030-43859-3_10

for design in architecture [2]. Comprehensibility of the formal process of shape generation was important to the exploration of design spaces. In architectural design combinations of shapes were often used to generate novel design ideas. The combinatorial approach using the set of geometric transformations move, scale, rotate and mirror was called shape grammar.

Shape grammar was applied to shape generation supporting creative processes [3, 4] and in designing variations of an underlying architectural theme [5, 6]. The mechanism, which expressed a grammar consisting of signs as shapes was called grammar translation. Based on this interpretive process, evolutionary computation was applied in grammar evolution.

Grammar evolution opened the potential for automated computation of shapes from underlying grammar [8–10]. Comprehensive investigations applying grammar evolution to architectural design were undertaken previously [11–14]. The design of grammars was a relevant consideration for the promotion of interactivity in evolutionary computation [15].

In 2001 Takagi's review of "Interactive Evolutionary Computation" [16] enhanced the discussion about human interaction within evolutionary computation. More research was conducted to gain knowledge about designer interaction in evolutionary computation [17, 18]. Evaluation of architectural solutions integrated aesthetics and functionality as criteria [19].

A case study combining quantitative and qualitative criteria to interactively evaluate design solutions in an evolutionary process was reported by Brintrup et al. [20]. Digital morphogenesis [21], performance-based modes of evaluation [22] and perception-based modes of evaluation [23] were investigated as main concepts for evolutionary design in architecture.

2 An Aesthetic-Based Fitness Measure

Aesthetics was a hard to evaluate criterion. It was highly subjective and varied in different experiences. One way to evaluate aesthetics was using a variety of aesthetic rules like Gestalt principles [24]. Those rules were based on cultural assumptions about aesthetics and restricted the solution space. Stepping away from pre-defined rules allowed the designer to define aesthetic rules implicitly through the evolutionary process.

A mechanism to provide direction to the shape generation process based on image input was explored. This initial image input was used as design reference. First, a similarity measure was calculated in comparison between generated shapes and input image. This similarity measure contributed to fitness evaluation.

Next, the calculated values of the similarity measure were stored. This process allowed to compare similarity values of the existing population as part of the evolutionary process. Similar shape solutions were omitted. Hence, similar shapes were not evaluated, so that novel shape solutions emerged from the population with higher probability. Moreover, the user did not evaluate similar shapes again in artificial evaluation. The four mechanisms that defined the aesthetic-based fitness measure shown in Fig. 1.

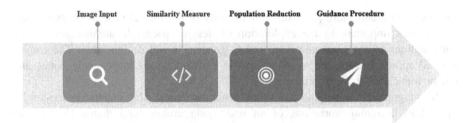

Fig. 1. Mechanisms for application of similarity measure

An image histogram defined the distribution of tonal values of the image [25]. During normalization of the image histogram each value of the histogram was divided by the total number of pixels to account for different image sizes. For computing the similarity measure, three distance metrics were chosen: Chebyshev, Euclidian and Manhattan distance of normalized histograms. Those distance metrics were used to compare the vectors of the normalized image histogram as displayed in Fig. 2.

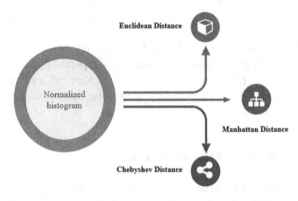

Fig. 2. Computing the similarity measure for aesthetic-based fitness measure

The combination of those three distance-metrics defined the similarity measure. The similarity measure calculated between image input and screen shots of the shape solutions contributed to the fitness function. This aspect of the investigated aesthetic-based fitness measure was named Shape Comparison.

3   A Framework for Guidance

In this section, the authors describe the framework for guidance of evolutionary design in architecture. The Genetic Programming (GP) process [7] was restricted by using strongly typed GP [26]. Figure 3 shows the modified GP process used in the framework for guidance. At the beginning of the evolutionary process, the population was initialized using Grow method. This method created short program trees for the shape grammar representation. The traditional Ramped Half-and-Half method integrating Full and Grow method or Uniform Initialisation were considered as alternative methods. The choice of initialisation method needed to be considered in connection with the shape representation. The amount of bloat generated by large program trees was crucial for the performance of GP. Seeding was considered as a possibility to add non-random solutions to the initial population [27, p. 41].

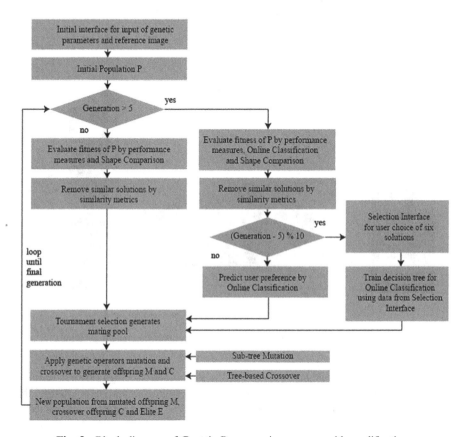

Fig. 3. Block diagram of Genetic Programming process with modifications

The initial population was used as input to the GP process and for the first five generations evaluated using performance measures and Shape Comparison. The Similarity measure was used to reduce the number of solutions after fitness evaluation. If the deviation of the similarity value from Shape Comparison was under a defined threshold, the solution was omitted for evaluation. Next, a set of solutions was selected from the current generation to define a mating pool as basis for the Tournament Selection procedure. Tournament Selection chose individuals for crossover using a direct comparison of individuals based on their fitness measure in repeated application.

Genetic operators were applied and generated two sets of offspring M and C. Here, M was generated from randomly chosen individuals of the current generation, which was computed by sub-tree mutation. C was result of subtree-swapping crossover using the Mating Pool created by Tournament Selection as input.

Both sets were combined with the best solutions of the generation preserved by Elitism (set E) to form the population for the next generation. The proportions of contributing solution sets M, C and E were defined by Mutation Rate and Crossover Rate specified by the user at the start of the process.

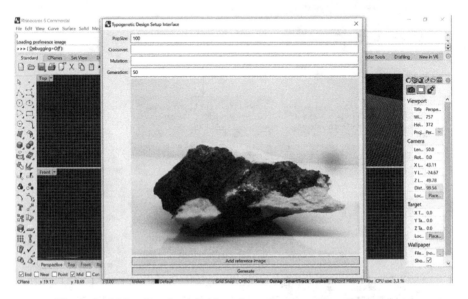

Fig. 4. Initial interface for definition of genetic parameters and image input

A first user interface allowed the user of the evolutionary design process to input those values along with population size and the number of generations. In addition, the reference image was collected as design reference. The initial interface for McNeel Rhinoceros CAD 6 using ETO library is shown in Fig. 4. This CAD software is currently state-of-the-art for design of complex shapes in architecture and frequently used in early design stages.

Until Generation 5, the fitness function combined performance measures and Shape Comparison. In all generations after Generation 5, the evaluation of the population used three mechanisms to calculate the fitness function: performance measures, Shape Comparison and Online Classification.

Online Classification used data generated by artificial selection of displayed shapes. A selection interface consisting of six shape displays with associated checkboxes was used to capture user preferences. The Selection Interface illustrated in Fig. 5 was shown to the user during every tenth generation as part of the shape design process.

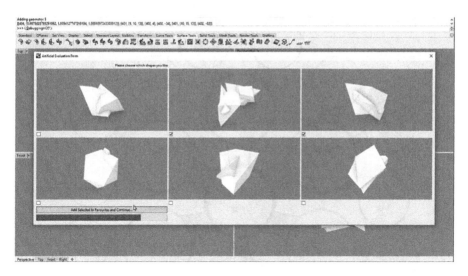

Fig. 5. Interface for artificial selection of shape solutions

The data used as input for the decision tree algorithm was extended continuously after every step of artificial selection. The decision tree was used to predict user preference of architectural shapes based on the representation data (genotype). The results contributed to the fitness function. This aesthetic measure modified the results of Shape Comparison, which were multiplied with the results predicted by decision tree algorithm.

Aesthetic-based fitness evaluation was facilitated using the initial interface for image input and the artificial selection interface to enable choice of solutions. Shape Comparison supported the designer by providing an aesthetic direction to grammar evolution. Online Classification predicted shapes likely to be preferred by the designer.

Besides the 3D shape grammar used in the case study, 2D shape grammar was an option for shape generation. Another possibility was the description of a House Grammar based on 3D shapes extended by building features as used in mass-customisation of houses [6].

The evaluation of multiple objectives provided the options of weighted-sum and non-dominant sorting. A detailed tutorial about multi-objective optimisation was provided by Emmerich and Deutz [28]. The last feature explored during as part of the framework was the similarity measure to reduce the number of solutions, which needed to be evaluated.

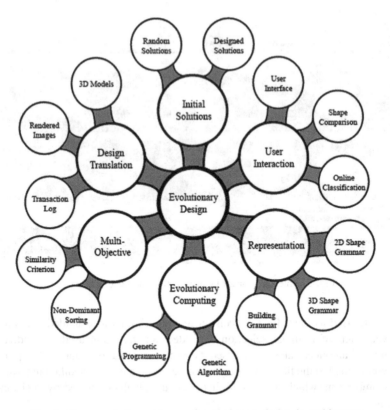

Fig. 6. Framework for guidance of evolutionary design in architecture

The different concepts contributing to the framework for guidance of evolutionary design in architecture are compiled in Fig. 6. The output data of the evolutionary process was a transaction log, which collected data about the GP run. Rendered images of the displayed and selected solutions were saved as part of the process, so that the user was able to review design decisions. The final output was a set of 3D shapes.

4 Experimental Setup

This section describes the experimental setup for the case study in Sect. 5. The case study applied the framework for guidance of evolutionary design in architecture to the design of a language museum in an architectural competition.

The shape grammar rules selected for the experiments were (a) translate, (b) rotate, (c) scale and (d) mirror. This simple ruleset allowed the designers to generate a wide range of shapes with a coherent aesthetic expression. The representation defined a combination of angled shapes.

Fitness measures used in experiment:

$$Fitness = Fitness_{Performance} + Fitness_{Aesthetics} \tag{1}$$

$$Fitness_{Aesthetics} = distance * predictor \tag{2}$$

$$Fitness_{Performance} = area/volume \tag{3}$$

The fitness function integrated the aesthetic-based fitness measure and the performance fitness in Eq. 1. The shape distance was measured as described in Sect. 2. This shape distance was modified by a predictor in Eq. 2. The predictor represented the prediction results of decision tree algorithm to predict shape preference by the designer.

The performance fitness function used area and volume as factors. The area of the building envelope was proportional to building cost, because façade area was the most expensive component of the building shape. The minimisation of the quotient area/volume was a measure for building sustainability. Compact shapes minimize the energy use of buildings. The built volume was proportional to the revenue generated by the building. Hence, the minimisation of the performance fitness function reduced energy use and building cost, while maximizing revenue as described in Eq. 3.

Table 1. GP parameters used during evolutionary design experiment

Parameter	Value
Population size	30
Generations	23
Crossover rate	0.6
Mutation rate	0.2
Tournament size	3
Matingpool size	3

During the case study, population size 25 and number of generations 23 were low to keep duration of the evolutionary process acceptable to the architects. Due to the time constraints of the architectural competition, the architects intended to keep design time for shape design under one day. Table 1 displays the GP parameters and values used during evolutionary design experiment.

The representation of shapes was defined by a strongly typed grammar [26] for two independent shapes and a set of shape grammar rules (move, rotate, mirror and scale). Repeated application of the shape grammar rules created complex shapes by combination of simple shapes.

In Fig. 7 the shape grammar process was illustrated to show the increasing complexity of shapes generated by the limited set of shape grammar rules.

Fig. 7. Generation of complex shapes using shape grammar rules

The basis for the grammar translation was the strongly typed grammar, which was described in Backus-Naur form as follows:

Grammar BNF-Form:

$$(\text{ComplexShape}) ::= (\text{Shape}), (\text{Shape}), (\text{Rules})$$
$$(\text{Shape}) ::= (\text{Point}), (\text{Point}), (\text{Point}), (\text{Point}), (\text{Point}),$$
$$(\text{Point}), (\text{Point}), (\text{Point}), (\text{Point}), (\text{Point})$$
$$(\text{Point}) ::= X, Y, Z$$
$$(\text{Rules}) ::= (\text{Rule}), (\text{Rule}), (\text{Rule}), (\text{Rule})$$
$$(\text{Rule}) ::= \text{Move}|\text{Rotate}|\text{Mirror}|\text{Scale}$$

The next section shows the case study, applying the framework for guidance of evolutionary design in architecture for an architectural competition.

5   Case Study

The competition brief provided a spatial program for a language museum in London. The chosen building shape translated the design idea into two merged shapes. The design idea was the intimacy of communication between two people. The exhibition design started below ground level with the history of language. The advancement of linguistic and communication systems in human culture was developed vertically towards the top of the building as part of the exhibition.

The unique shapes generated by the evolutionary process allowed the designers to compare shape aesthetics close to real-time. In parallel to the evolutionary design process, other forms of communication (like sketching, discussion and design reviews) were used by the architects. Grammar evolution enhanced the communication and collaboration during shape design. The process allowed the designers to review a vast range of shape expressions and discuss the aesthetic impact.

The architectural shapes displayed in Fig. 8 were identified as a suitable geometric expression for the design case at hand, reflecting the symbolic idea of a gem in their multi-facetted geometry. The individual expression of the museum was generated as architectural shape with an acceptable level of complexity to ensure buildability.

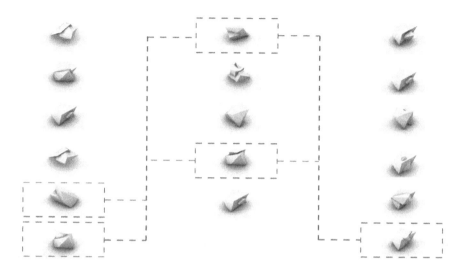

Fig. 8. Development of shapes generated during case study

The decision path in Fig. 8 shows user choices during shape design. Due to site configuration, elongated shapes were preferred by the designers. The separation between exhibition and teaching/administration spaces in the functional program led to a clearly articulated intersection between the two building masses.

The language museum was designed as an expressive landmark in central London. In response to the brief, the proposed design connected exhibition and learning spaces internally, while showing a legible differentiation on the building surface. The large exhibition and gallery spaces were separated from the subordinated learning, teaching and administration spaces. This key design motive emerged during the evolutionary shape design process.

After shape design, the 3D model was positioned as an elongated body at the edge of the building site. A protected outside space was created between the proposed design and the neighbouring buildings. After introducing levels to the 3D model, floor plans were developed to negotiate the different functions demanded in the brief. Some further adjustment to the geometry were necessary to generate the qualities of internal spaces imagined by the designers.

The application of the framework for guidance of evolutionary design in architecture provided the architectural shape as basis for the further design steps illustrated in Fig. 9. Furthermore, the framework was augmented collaborative decision-making.

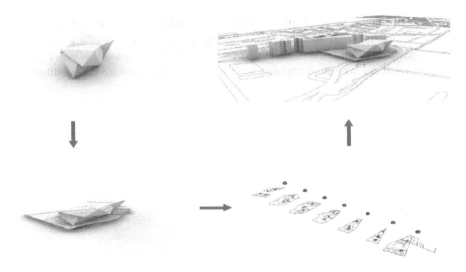

Fig. 9. Design process from shape generation to competition entry

The framework was used in the design team to discuss the aesthetic implication of different architectural shapes. During this process, spatial qualities and aesthetic expression of design options observed from different perspectives were negotiated easily.

After reconsidering the site development and making additional changes to the façade geometry, the design proposal was visualized. The resulting aerial of the architectural competition and an architectural rendering is presented in Fig. 10.

Fig. 10. Architectural rendering illustrating the architectural competition

During the design of the language museum, the contributing architects reduced the design time for the building shape to under one hour. One of the architects stated that designing similar shapes for other projects without the evolutionary design process took approximately three days. Both architects agreed on the final shape regarding form and function. This statement confirmed the value of applying the framework for guidance of evolutionary design in architecture.

6 Discussion

In architectural design, optimisation processes for energy consumption, material use and building cost were usually applied after shape design, if at all. Calculating building costs and comparing them to budget constraints in early design reduced the number of designs, which need to be planned to a higher level of detail. Optimisation of shape geometry in early design could additionally increase energy, material and cost efficiency. The unrecognized impact of envelope geometry on building performance was one motivation to investigate the framework for guidance of evolutionary design in architecture.

Adding design references by images, one of the preferred ways of architects to creatively interact in design was integrated into evolutionary design. This mechanism engaged the users at the beginning of the evolutionary design process. Therefore, the interest of the designers in the generated architectural shapes was increased. The input capabilities should be tested using sketches. Additionally, other fitness mechanisms [29] could contribute to the reduction of user effort.

Increasing adaptivity and flexibility of the design process could be achieved by using the framework for meta design [15] of grammar representations. Genetic programming [7] and strongly typed grammar [26] could be applied to evolve grammar rules. Using a more general representation for specific design cases, for examples hospitals, schools or houses could be used to extract design rules from the user input data.

An evolutionary design approach for floorplans, as displayed in the context of generic shape grammar for the Palladian Villa, Malagueira house, and prairie house [30] would be a suitable approach to collect data about functional considerations for floor plan layout. The extraction of aesthetic rules as design principles would be a possible application of the framework. This process might support architects in reflecting actively on their aesthetic perspective.

The author's research suggested that the use of a broad interactive approach in shape design reduced the effort of shape evaluation by omitting many shape options using similarity measure. Fitness prediction by using the results generated by evaluating large shape datasets could reduce computational cost.

The overall computation time of shape solutions was reduced by elimination of similar solutions. During initial generations - before interactive evaluation - the guidance by Shape Comparison allowed the evolutionary system to advance the shape generation, so that the designer started artificial selection from an advanced position.

As a result, the designer was not confronted with random solutions at the beginning of the evolutionary design process.

7 Conclusions

In conclusion, the authors presented a novel approach to interactive shape design, using an aesthetic-based fitness measure in reference to an initial image input. This mechanism increased the efficiency during exploration of a design space for architectural shape design. The visual feedback during the evolutionary process increased the engagement of the designers. In addition, the framework was valuable for discussion of aesthetic features of architectural solutions. The framework supported the collaborative decision making during the case study.

Integrating performance-based modes of evaluation [22] and perception-based modes of evaluation [23] in a framework was a step toward performance-based creative systems. The investigated framework supported the designers during the case study in aesthetic decision-making as part of the creative process. Exploring a design space after deciding on a shared grammar representation stimulated the designers' imagination. The provided case study showcased the application potential of the framework to interactively explore design spaces. Those design spaces were previously explored either automatically with low engagement of the architect or manually with high user effort.

8 Recommendations

Further testing of the guidance procedures in user experiments and usability studies will increase the efficiency of the framework for guidance of evolutionary design in architecture. Collection of additional qualitative data about user behaviour might lead to a better understanding of designer needs regarding evolutionary design. A potential study in this area could include questions about the engagement of the designer during evolutionary design and reflective questions on their own perception of the evolutionary design process. Also, the impact of the aesthetic-based guidance mechanism on effects of designer fatigue and user fatigue could be evaluated in empirical studies.

Additional case studies with different representations will be undertaken. From the authors' perspective, an application for mass-customisation of 3D printed architectural components would be a possible area of application. Reducing the complexity of 3D printed structures to the connection details minimized waste in 3D printing and rationalized the application of additive manufacturing technology [31]. Using serial products for structural members and façade panels reduced production time and cost.

At the same time, 3D printed structural nodes enabled geometric complexity of architectural designs [31]. The framework for guidance of evolutionary design in architecture will be applied on multiple levels of scale in this application case. On one hand, the design of the 3D printed structural components will be enhanced by utilizing the framework and linking it to manufacturing as displayed in Fig. 11.

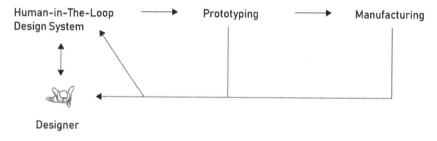

Fig. 11. Linking the framework for evolutionary design in architecture to manufacturing

On the other hand, architectural shapes generated by the framework will be rationalized for mass-customization. One option for integrating the framework for evolutionary design in architecture into a parametric design environment would be Grasshopper. This option would allow to link the framework to the available simulation capabilities. In this way, virtual prototypes of architectural shapes could be evaluated interactively [32]. Finally, the use of the aesthetics-based guidance mechanism in the parametric context would extend the generative design paradigm described by Frazer [33] by adding an aesthetic system for exploration of large datasets of architectural shapes.

References

1. McCormack, J., Dorin, A., Innocent, T.: Generative design: a paradigm for design research. In: Redmond, J., De Bono, A. (eds.) Proceedings of the Futureground Design Research Society International Conference, vol. 2, 1 edn. Monash University (2004)
2. Singh, V., Gu, N.: Towards an integrated generative design framework. Des. Stud. **33**(2), 185–207 (2012)
3. Stiny, G., Gips, J.: Shape grammars and the generative specification of painting and sculpture. In: IFIP Congress (2), vol. 2, no. 3 (1971)
4. Knight, T.W.: Designing a shape grammar. In: Gero, J.S., Sudweeks, F. (eds.) Artificial Intelligence in Design 1998, pp. 499–516. Springer, Dordrecht (1998). https://doi.org/10.1007/978-94-011-5121-4_26
5. Flemming, U.: More than the sum of parts: the grammar of Queen Anne houses. Environ. Plan. B: Plan. Des. **14**(3), 323–350 (1987)
6. Duarte, J.P.: Customizing mass housing: a discursive grammar for Siza's Malagueira houses (2001)
7. Koza, J.R.: Genetic programming - a paradigm for genetically breeding populations of computer programs to solve problems. Stanford University (1990)
8. al-Rifaie, M.M., Ursyn, A., Zimmer, R., Javid, M.A.J.: On symmetry, aesthetics and quantifying symmetrical complexity. In: Correia, J., Ciesielski, V., Liapis, A. (eds.) EvoMUSART 2017. LNCS, vol. 10198, pp. 17–32. Springer, Cham (2017). https://doi.org/10.1007/978-3-319-55750-2_2
9. Muehlbauer, M., Burry, J., Song, A.: Automated shape design by grammatical evolution. In: Correia, J., Ciesielski, V., Liapis, A. (eds.) EvoMUSART 2017. LNCS, vol. 10198, pp. 217–229. Springer, Cham (2017). https://doi.org/10.1007/978-3-319-55750-2_15

10. Frazer, J.: An Evolutionary Architecture. Architectural Association, London (1995)
11. Janssen, P.: A design method and computational architecture for generating and evolving building designs. The Hong Kong Polytechnic University (2004)
12. O'Neill, M., et al.: Evolutionary design using grammatical evolution and shape grammars: designing a shelter. Int. J. Des. Eng. **3**(1), 4–24 (2010)
13. Muehlbauer, M., Song, A., Burry, J.: Towards intelligent control in generative design. In: Cagdas, G., Ozkar, M., Gul, L.F., Gurer, E. (eds.) Future Trajectories of Computation in Design, Proceedings of CAAD Futures 2017, Istanbul, pp. 630–647 (2017)
14. Byrne, J., McDermott, J., Galván-López, E., O'Neill, M.: Implementing an intuitive mutation operator for interactive evolutionary 3D design. In: IEEE Congress on Evolutionary Computation, pp. 1–7. IEEE (2010)
15. McDermott, J., et al.: String-rewriting grammars for evolutionary architectural design. Environ. Plan. B: Plan. Des. **39**(4), 713–731 (2012)
16. Takagi, H.: Interactive evolutionary computation: fusion of the capabilities of EC optimization and human evaluation. Proc. IEEE **89**(9), 1275–1296 (2001)
17. Lehman, J., Clune, J., Misevic, D.: The surprising creativity of digital evolution. In: Artificial Life Conference Proceedings, pp. 55–56. MIT Press, Cambridge (2018)
18. Kowaliw, T., Dorin, A., McCormack, J.: Promoting creative design in interactive evolutionary computation. IEEE Trans. Evol. Comput. **16**(4), 523 (2012)
19. Schumacher, P.: The Autopoiesis of Architecture. Wiley, Chichester (2011–2012)
20. Brintrup, A.M., Ramsden, J., Takagi, H., Tiwari, A.: Ergonomic chair design by fusing qualitative and quantitative criteria using interactive genetic algorithms. IEEE Trans. Evol. Comput. **12**(3), 343–354 (2008)
21. Roudavski, S.: Towards morphogenesis in architecture. Int. J. Arch. Comput. **7**(3), 345–374 (2009). https://doi.org/10.1260/147807709789621266
22. Shea, K., Aish, R., Gourtovaia, M.: Towards integrated performance-driven generative design tools. Autom. Constr. **14**(2), 253–264 (2005)
23. Gips, J., Stiny, G.: Artificial intelligence and aesthetics. In: IJCAI, vol. 75, pp. 907–911 (1975)
24. Muehlbauer, M.: Towards typogenetic tools for generative urban aesthetics. In: Biloria, N. (ed.) Smart and Sustainable Built Environment, vol. 7, no. 1, pp. 20–32 (2018). https://doi.org/10.1108/sasbe-10-2017-0052
25. Krutsch, R., Tenorio, D.: Histogram Equalization. Freescale Semiconductor, Document Number AN4318, Application Note (2011)
26. Montana, D.: Strongly typed genetic programming. Evol. Comput. **3**(2), 199–230 (1995). https://doi.org/10.1162/evco.1995.3.2.199
27. Poli, R., Langdon, W.B., McPhee, N.F., Koza, J.R.: A Field Guide to Genetic Programming. Lulu.com (2008)
28. Emmerich, M.T.M., Deutz, A.H.: A tutorial on multiobjective optimization: fundamentals and evolutionary methods. Nat. Comput. **17**(3), 585–609 (2018). https://doi.org/10.1007/s11047-018-9685-y
29. Johnson, C.G.: Fitness in evolutionary art and music: what has been used and what could be used? In: Machado, P., Romero, J., Carballal, A. (eds.) EvoMUSART 2012. LNCS, vol. 7247, pp. 129–140. Springer, Heidelberg (2012). https://doi.org/10.1007/978-3-642-29142-5_12
30. Benrós, D., Hanna, S., Duarte, J.P.: A generic shape grammar for the Palladian Villa, Malagueira house, and Prairie house. In: Gero, J.S. (ed.) Design Computing and Cognition 2012, pp. 321–340. Springer, Dordrecht (2014). https://doi.org/10.1007/978-94-017-9112-0_18

31. Crolla, K., Williams, N., Muehlbauer, M., Burry, J.: Smartnodes pavilion - towards custom-optimized nodes applications in construction. In: Janssen, P., Loh, P., Raonic, A., Schnabel, M. (eds.) Protocols, Flows, and Glitches – Proceedings of the 22nd CAADRIA Conference (2017)
32. Burry, M., Burry, J.: Prototyping for Architects, p. 53. Thames and Hudson Ltd., London (2016)
33. Frazer, J.: Parametric computation: history and future. Arch. Des. **86**(2), 18–23 (2016)

Objective Evaluation of Tonal Fitness for Chord Progressions Using the Tonal Interval Space

María Navarro-Cáceres[1](✉), Marcelo Caetano[2,3], and Gilberto Bernardes[4]

[1] Department of Computer Sciences, University of Salamanca,
Pza de los Caídos, s/n., 37007 Salamanca, Spain
maria90@usal.es

[2] CIRMMT, Schulich School of Music, McGill University, Montreal, QC, Canada
marcelo.caetano@mcgill.ca

[3] Aix-Marseille Univ, CNRS, PRISM "Perception, Representation,
Image, Sound, Music", Marseille, France
marcelo.caetano@prism.cnrs.fr

[4] INESC TEC, Faculty of Engineering, University of Porto, Porto, Portugal
gba@fe.up.pt

Abstract. Chord progressions are core elements of Western tonal harmony regulated by multiple theoretical and perceptual principles. Ideally, objective measures to evaluate chord progressions should reflect their *tonal fitness*. In this work, we propose an objective measure of the fitness of a chord progression within the Western tonal context computed in the Tonal Interval Space, where distances capture tonal music principles. The measure considers four parameters, namely tonal pitch distance, consonance, hierarchical tension and voice leading between the chords in the progression. We performed a listening test to perceptually assess the proposed *tonal fitness* measure across different chord progressions, and compared the results with existing related models. The perceptual rating results show that our objective measure improves the estimation of a chord progression's *tonal fitness* in comparison with existing models.

Keywords: Chord progression · Hierarchical tension · Tonal Interval Space · Melodic attraction · Consonance

1 Introduction

Chords are fundamental elements of Western tonal music. The vertical construction of chords and its horizontal motion, known as chord progressions, have been the subject of several theories over the past decades [15, 17, 18]. Most theoretical analyses of chord progressions focus on particular elements among the multidimensional principles regulating chord progressions, such as consonance, musical tension and voice leading. Among these theories, we can highlight those that express tonal pitch relations in topological spaces [8, 17] aiming to capture the

© Springer Nature Switzerland AG 2020
J. Romero et al. (Eds.): EvoMUSART 2020, LNCS 12103, pp. 150–164, 2020.
https://doi.org/10.1007/978-3-030-43859-3_11

sense of proximity between chords in Western tonal music. In the aforementioned theories, chords in a progression are addressed both linearly and hierarchically, thus highlighting the importance between consecutive chords and their function across multiple hierarchies. In this paper, the objective measure of how well a given chord fits a progression is called *"tonal fitness"* [1].

Some authors propose to measure the *tonal fitness* of the next chord in a progression using the previous chord as reference, therefore, only considering the linear dimension of a chord progression. Quick *et al.* [13] present a grammar with probabilities based on a previous analysis of Bach pieces. Woolhouse *et al.* [21] propose the pitch attraction model, which evaluates one chord according to the previous one. Callender *et al.* [6] present a generalized space for chord representation to evaluate two consecutive chords. There are proposals which evaluate chords by considering more than a single previous chord in the progression [7,12,20]. These works analyze style-specific music corpora to statistically extrapolate long-term features, such as typical movements of tonal functions or common melodic sequences. However, the resulting models do not capture long-term dependencies such as phrase structures. The evaluation of the fitness of the chords within the hierarchical dimension of music structure requires a hierarchical analysis.

To address this limitation, several authors propose models in which the tonal properties of a chord are measured by considering not only on the previous chords but also their hierarchical relationships, typically represented as a tree structure [8]. Bernstein draws a basic relationship between music and Chomsky's formal grammars [5]. Schenker [17] proposes a hierarchical analysis to reduce the musical surface to tonal functions, which lacks a comprehensive computational formulation. Steedman [19] designs a context-free grammar to model Blues progressions. Rohrmeier [16] presents a grammar to generate structures for tonal music based on hierarchical trees that can capture different tonal hierarchies. Lerdahl's [8] proposal to measure tonal tension and melodic attraction of a chord progression from linear and hierarchical structures of a musical phrase is one of the most influential models to date. Lerdahl adopts four tonal indicators computed manually from chord progressions: *stability*, *consonance*, *hierarchical tension*, and *voice leading*.

In our work, we depart from the conceptual basis of Lerdahl's model to computationally measure the *tonal fitness* of a chord within a progression. Tonal pitch indicators inspired by Lerdahl's model are computed in our work using the Tonal Interval Space (TIS) by Bernardes *et al.* [4], where hierarchical tonal pitch relations are expressed as distances. Use of the discrete Fourier transform (DFT) in the TIS to calculate the distances results in a computationally efficient representation. Besides, this representation automatically defines the regional (or

---

[1] Lerdahl's and Farbood's proposals use the term *tonal tension* to refer to how a chord is hierarchically related to the rest of the chords in a progression. Following this concept, we will use "tonal fitness" to capture how well a chord reflects tonal properties in the context of the chord progression, according to the Western tonal rules. It therefore includes Lerdahl's concept of tonal tension.

key) space of a given chord progression, in contrast with Lerdahl's model, which requires manual definition. Therefore, our work results in a more flexible computational analysis framework which also broadens the scope of target applications and users by excluding the need for prior musical knowledge or annotations on the analyzed musical data.

Furthermore, our work addresses both linear and hierarchical long-term dependencies of chord progressions, which also extend the manually-driven Lerdahl's proposal with an automatic computational method for representing musical hierarchies from a given chord progression as tree structures. In short, stemming from state-of-the-art models, our approach not only fosters a computationally efficient framework for the analysis of chord progressions' *tonal fitness*, but also aims to improve the accuracy of tonal indicators by adopting the perceptually-inspired TIS. Ultimately, we believe that our contribution can support and enable new tools for the automatic analysis and generation of hierarchically-aware tonal chord progressions, whose lack of hierarchical structure has been a long identified problem in the field of generative music [10].

To validate our measure, we conducted a listening test with chord progressions from different tonalities. We compare our currently proposed measure with a previous work [11] whose measure only considers the linear dimension of the chord progressions, and with Lerdahl's [8] measure. Comparison with a previous work allows to analyze how the hierarchical and linear dimensions can influence the subjective ratings in the same representation space, whereas comparison with Lerdahl's measure aims to assess how the representation space can affect the measurement of the *tonal fitness*.

The remainder of this paper is structured as follows. Section 2.1 describes the TIS and its features. Section 2 explains the theoretical paradigm that influences the formalization of the measure. Section 3 describes the measure to calculate the *tonal fitness* of the chords following linear and hierarchical criteria. Section 4 details the experiment, and Sect. 5 describes the results obtained and the discussion. Finally, Sect. 6 presents the conclusions and the future work.

2   Theoretical Background About Chord Progressions

2.1   Tonal Interval Space

Fourier analysis has been recently used to explore tonal pitch relations [2, 4, 14]. Bernardes *et al.* [4] proposed to calculate the weighted DFT of a chroma vector $C(n)$ to obtain Tonal Interval Vectors (TIVs). The TIVs can be used to represent different tonal pitch hierarchies, such as pitch classes, intervals, chords, and keys.

One of the most important properties of the TIS is that distances among different pitch configurations represented by their TIV reflect musical attributes. Each TIV is interpreted as vector $T(k)$ with $k = 1, \cdots, M = 6$, where M is the number of components considered from the DFT. The Euclidean distance between $T_1(k)$ and $T_2(k)$ is given by

$$\rho(T_1, T_2) = \sqrt{\sum_{k=1}^{M} |T_1(k) - T_2(k)|^2}. \tag{1}$$

Perceptually, similar vectors result in smaller distances than dissimilar ones. The inner product between $T_1(k)$ and $T_2(k)$ is

$$T_1 \cdot T_2 = \sum_{k=1}^{M} T_1(k) \overline{T_2(k)}, \tag{2}$$

where $\overline{T_2(k)}$ is the complex conjugate of $T_2(k)$. Equation (2) yields higher values for perceptually similar vectors than dissimilar ones. The angle θ between T_1 and T_2 can be calculated from Eq. (2) as

$$\theta(T_1, T_2) = \arccos\left(\frac{T_1 \cdot T_2}{\|T_1\|\|T_2\|}\right), \tag{3}$$

where $\|T_1\|$ denotes the magnitude of T_1 calculated as the distance $\rho(T_1, T_0)$ between T_1 and the center of the space T_0 using Eq. (1). Equation (3) results in smaller angles to indicate a higher degree of similarity.

Equations (1) to (3) can be used to compute distances between pitch configurations of the same level (e.g., the distance between two chords) or across levels (e.g., the distance between a chord and a key). For example, the distance between two chords captures their tonal relatedness, the distance between a chord and a key measures the level of membership of the chord to the key, and the distance of a given chord to the three categorical harmonic functions (represented in the space as the triads of the tonic, subdominant and dominant degrees) measures how well the chord fits the categorical harmonic functions. All these properties are used in the proposed measure (Sect. 3) to capture the suitability of the chord in the context of a progression.

2.2 Lerdahl's Theory to Model the Tonal Fitness of Chord Progressions

Some authors have proposed models of structural dependencies of chord progressions [8,16,17]. In particular, Lerdahl [8] proposed a measure $L(T_i, P)$ that assesses the tonal tension and the melodic attraction of a chord T_i in a progression P. Lerdahl's model explores the role a chord plays within the hierarchical structure of a chord progression by considering four elements, namely, *tonal pitch distance, surface dissonance, voice leading,* and *hierarchical structure.*

Tonal pitch distance captures the proximity of chords using an algebraic representation. Lerdahl proposes a space that measures chord distances using a lookup table, in which non-common tones between chords, key distances, and the interval distance among the chords root are considered [9].

Surface dissonance measures the psychoacoustic dissonance of chords in a progression from the interaction among the vertical component notes of each chord. Surface dissonance results from the combination of three factors: "scale degree", which considers the component scale degrees in a given chord; "inversion", which accounts for the chord's bass note and on its subsequent elaboration as root position or inversion; and "nonharmonic tone", which inspects the existence of tones outside the chord function.

Voice leading captures the melodic attraction between consecutive chords per voice. Voice leading is a horizontal measure that estimates the fitness of each note in a voice according to the previous note. Theoretically, the number of semitones, the stability of the notes, and the overall fitness between the chords can affect the evaluation of the voices [9,14]. Lerdahl [9] states that the stability of each note is related with the distance to the tonal center or key. The closer to the key, the more stable the note is.

Hierarchical structure captures the multiple levels of the musical structure by adopting tree-based structures driven from functional harmonic categories. Lerdahl's hierarchical approach analyzes the hierarchical structure through the tonal tension measure, which can be divided into local tension and inherent tension to know how the chords are hierarchically related. Local tension measures the distance between the chord we want to evaluate and the chord immediately above it in the tree, called parent chord. Likewise, inherent tension considers all the distances between the parent chord and all the chords above.

2.3   Rohrmeier's Hierarchical Tree Structure

Rohrmeier's computational approach to encode the hierarchy of chord progression is detailed, towards the definition of an evaluation scheme for hierarchical tonal fitness. Rohrmeier applies principles from theories focusing on the hierarchical dimension of music using a binary grammar that respects tonal rules to generate chord progressions and explicitly designed to be computationally feasible and testable [16]. His hierarchical model of tonal music relies on four categories, of which the functional level is of relevance here. Consequently, we apply this generative grammar to our work to measure the hierarchical structure.

According to Rohrmeier, the rules in Eqs. (4a) to (4g) characterize the core behavior of a grammar which represents the functional regions in a chord progression. The rules contain two kinds of symbols: *non-terminal symbols* represented by capital letters and *terminal symbols* represented by lower-case letters. The *non-terminal symbols* represent functional regions such as the tonic region TR, the dominant region DR, the subdominant region SR, and any of the previous regions XR. Non-terminal symbols can be expanded into different harmonic regions following the rules in Eqs. (4a) to (4d) or into terminal symbols following Eqs. (4e) to (4g). The *terminal symbols* t, d, s represent tonic, dominant, or subdominant chords in the sequence respectively, and cannot be further expanded.

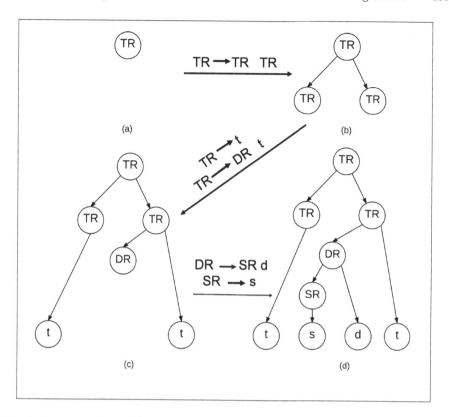

Fig. 1. Representation of the construction of a hierarchical structure for a three-chord progression.

$$TR \longrightarrow TR \quad DR \qquad (4a)$$

$$TR \longrightarrow DR \quad t \qquad (4b)$$

$$DR \longrightarrow SR \quad d \qquad (4c)$$

$$XR \longrightarrow XR; \quad XR \quad \forall \, \text{non-terminal} \qquad (4d)$$

$$TR \longrightarrow t \qquad (4e)$$

$$DR \longrightarrow d \qquad (4f)$$

$$SR \longrightarrow s \qquad (4g)$$

Figure 1 shows an example of how to construct a tree structure. The first node in Fig. 1(a) is a tonic region TR, which is responsible for defining the key of the chord progression. Firstly, we apply Rule (4d) for the tonic region. Then, the first child TR applies the Rule (4e) and the second child applies (4b) (Fig. 1(c)). The final Fig. 1(d) is obtained by applying Rules (4c) and (4g).

3   Measuring the Tonal Fitness of a Chord Progression

The properties of the TIS derived from mathematical measures such as the Euclidean distance or the angle between TIVs are the main indicators used to measure the tonal fitness of a pitch configuration [3]. In the present work, we aim to create a measure that captures the tonal fitness of a chord through its musical properties. Following Lerdahl's proposal, we create a measure in Eq. (5), where each term corresponds to the codification of an element proposed in Sect. 2 but encoded in the TIS: *tonal pitch distance* (δ), *surface dissonance* (c), *voice leading* (m), and *hierarchical structure* (h):

$$M(T_i, P) = k_1\delta(T_i, P) + k_2c(T_i) + k_3m(T_i, P) + k_4h(T_i, P), \qquad (5)$$

where T_i is the ith-chord of the progression P and k_j are constants that represent the weights of each parameter. The following subsections will detail how the four items are encoded in the TIS so that we can measure them mathematically.

3.1   Tonal Pitch Distances

The measure δ encodes the tonal pitch distance between two consecutive chords. According to Sect. 3.1, δ measures three musical properties: distance to the previous chord in the sequence, distance to the key, and distance to the tonal function. In Eq. (6), δ is decomposed as

$$\delta(T, P) = \rho(T_i, T_{i-1}) + \theta(T_i, T_{key}) + \theta(T_i - T_{key}, T_f), \qquad (6)$$

where $\rho(T_i, T_{i-1})$ uses Eq. (1) and θ uses Eq. (3). $\theta(T_i, T_{key})$ measures the degree of membership of T_i to the main key of the full chord progression and $\theta(T_i - T_{key}, T_f)$ measures the similarity to the tonal function given by the tree previously built.

In the TIS, chords musically close to the key have small distances between the key configuration and the chord configurations. The key is represented by T_{key} obtained as the TIV of the chroma vector $C_{key}(n)$ that contains all the notes of the scale corresponding to the key. Then, Eq. (3) estimates the angle between the chord T_i and the center of the key T_{key}.

To measure the proximity of one chord to a tonal function in the TIS, we check if the chord T_i is aligned with the tonic I, subdominant IV, or dominant V by using the angle measure proposed in Eq. (3). $T_i - T_{key}$ is the chord using T_{key} as reference and T_f is a vector representing one of the harmonic functions I, IV, and V also referenced by T_{key}. We aim to minimize $\theta(T_i, T_{key}, T_f)$ using T_f following the harmonic sequence given by the tree built in the previous step to ensure T_i is aligned with one of them.

3.2   Consonance

In the TIS [3], the norm $\|T_i\|$ measures the consonance c of the chord represented by T_i. If c results in large values of $\|T_i\|$, the chord is very consonant. Therefore, we aim to maximize $c(T_i)$.

3.3 Voice Leading

The measure m represents the melodic attraction of two consecutive chords. Lerdahl proposes three factors to model the melodic attraction, which are the perceptual distance between the chords, the number of semitones between the notes and the stability of the notes (how much they attract the rest of the notes). Following this, the voice leading measure between T_i and T_{i-1} has been encoded as

$$m(T_i, P) = \frac{\sum_{l=1}^{3} v(n_{l_{i-1}}, n_{l_i})}{\rho(T_i, T_{i-1})}, \tag{7}$$

where ρ is the Euclidean distance from Eq. (1) which captures the perceptual distance between the chords and v is a measure of voice-leading for each voice l produced between the note n_{l_i} of the present chord and the note $n_{l_{i-1}}$ of the previous chord. To calculate the stability of the notes of a chord in a progression, we measure the distance of its corresponding pitch class to the key in the TIS. The number of semitones and the distance to the key are calculated as

$$v(n_{l_i}, n_{l_{i-1}}) = \rho(T_{n_{l_i}}, T_{key})e^{0.05s} \tag{8}$$

where s is the number of semitones between n_{l_i} and $n_{l_{i-1}}$.

3.4 Hierarchical Tension

The value of h represents the tonal tension concept following Lerdahl's theory, which considers distances between chords that are closer in a tree structure modeled for a progression. The tension related to the hierarchical structure of the progression is calculated as

$$h(T_i, P) = \rho(T_i, T_j) + \frac{\sum_{k=j}^{N} \rho(T_i, T_k)}{N}, \tag{9}$$

where N is the number of chords in the tree and T_j is the parent chord of T_i and T_k is the parent chord of T_{k-1} following the tree structure of the chord progression P. Here, ρ is the Euclidean distance between two chord codifications calculated with Eq. (1).

Rohrmeier *et al.* [16] proposes a generative grammar to create hierarchical structures following tonal music principles. However, Rohrmeier's work is not oriented to generate chord progressions computationally based on his grammar. Given a progression, we need to extract one tree that represents the hierarchy between the chords in the progression. Therefore, it is necessary to implement a method that calculates the node in the tree corresponding to the specific chord considered by inversely applying the rules of Rohrmeier's proposal [16]. Firstly, the leaf chords are always replaced by t, s or d. To decide which harmonic function is aligned with each chord, we calculate the angle between the tonal function and the chord in the TIS. In a second step, we apply the grammar rules inversely following a hierarchy according with three criteria:

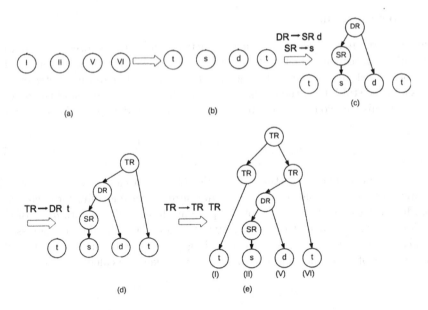

Fig. 2. Process to obtain a tree structure from a chord progression.

- To avoid unfeasible trees, select first the element 's', which appears in a lower number of rules.
- To avoid unfeasible trees, try to build the tree from the inner leaves and to connect them with the outer ones.
- To avoid trees with too many nodes, always try to apply a rule which implies a greater number of elements, if possible. In case we have several rules that accomplishes this criterion, select one randomly.

To clarify this process, a simple example considering the tree of Fig. 1 is illustrated in Fig. 2. In the first stage, the chords are replaced by 't', 's' or 'd', according to their alignment with the harmonic functions. Secondly, as we have an 's' element and this can be connected to the 'd' element, so we apply Rule (4c) (Fig. 2(c)). Now, we can connect the first 't' element, or the last 't' element. Randomly, we selected the final 't' and applied inversely Rule (4b). Finally, we connected the first 't' with Rule (4e) and Rule (4d) (Fig. 2(d)).

With the hierarchical tree, the tension can be now calculated. However, we still need to know the importance of a chord in the hierarchical structure and consequently their local tension and inherent tension. To calculate their tension, we need to know which chords dominate (have more tension) the other chords.

The hierarchy of "dominated" chords is represented by replacing each node of the tree constructed previously with the most tense chord, selected from its children. We implemented a method to obtain a tree with the "dominant" chords of each level, and therefore, to be able to calculate the local and inherent tensions. Firstly, the leaf nodes are replaced by the specific chords. The parent nodes with only one child automatically represent the leaf chord. The parents with

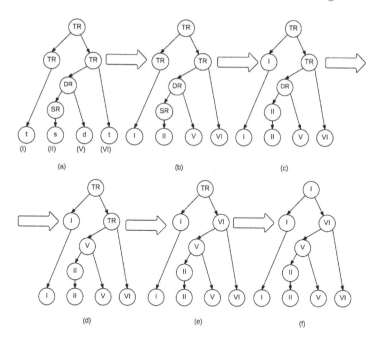

Fig. 3. Visualization of a tree structure with specific chords to measure the tension.

children representing different tonal regions are automatically represented by the child with the same harmonic function that the parent. Finally, the parent with children of the same tonal regions is replaced by the most stable chord according to the measure described in Sect. 3.1. A simple example considering the tree of Fig. 1 is illustrated in Fig. 3. Step a represents the initial tree and the chord progression. Step b replaces the terminal by the degrees. Step c replaces the SR and TR nodes with the only child they have. Step d replaces the DR with V. Similarly, step e replaces TR with the tonic chord V and step f chooses the most stable chord according to the tonal distance proposed in the previous section, selecting I as the parent node.

4   Evaluation

The evaluation aims to demonstrate that M reflects the *tonal fitness* of a chord in a progression. Firstly, we need to evaluate $M(T_i, P)$ in optimal conditions. That means we need to find the best values for weights k_j in Eq. (5) by applying cross validation. In a second step, we validate $M(T_i, P)$ by comparing subjective ratings collected by a listening test and the objective measure for different chord progressions.

Two premises were considered to create the listening test. Firstly, the measure M should be able to capture tonal fitness for both *fit* and *unfit* chords. Thus chords with a high level of *tonal fitness* would have low values of M and vice-versa. Secondly, M should reflect their tonal fitness independently of the main key and of the hierarchical tree. Additionally, we have to consider that the evaluation of harmony could be influenced by multiple factors. To avoid external elements that can bias the subjective, the listening test contains controlled chord progressions in a basic musical context with specific and short chord progressions.

The listening test presented a sequence of three-note chords and asked the listener to rate how well the last chord fits the progression. Note that the three-note chords can be triads or contain non-harmonic tones. To demonstrate that $M(T_i, P)$ reflects the *tonal fitness* of chords, we selected triad chords randomly but with objective ratings that sample the function M from low values (chords with a high *tonal fitness*) to high values. The first chord in each progression was always the tonic in root position to establish the key [1] because the tonic determines the tonal basis of the music. Additionally, the root position triads have a firmer sense of tonal centering, resulting in the difference in pitch between the major and minor modes. The second chord is always different for each key to represent different harmonic functions. Additionally, to demonstrate its independence of tree structures and keys, the chord progressions were classified in four groups, two for a tree with the sequence Tonic-Dominant-Tonic, in G major and C minor, and two for a tree with the sequence Tonic-Subdominant-Dominant, in C minor and G minor. Both tree structures with some examples of chord progressions for each tree are shown in Fig. 4.

The listening test is online[2] and consists of four playlists with the chord progressions presented randomly. Each chord progression can be played multiple times before assessing it. The listeners were asked to evaluate how well the third chord follows the first and second using the following ratings: very good $(+2)$, good $(+1)$, fair (0), bad (-1), or very bad (-2). In total, 48 people took the test, among which ten declared no musical training, nineteen considered themselves amateurs, and twenty were professional musicians.

We expect the objective measure $M(T_i, P)$ to correlate well with the ratings from the subjective evaluation and reflect the *tonal fitness* of each chord because $M(T_i, P)$ includes information from the linear and hierarchical dimensions. The sum of the distances in the hierarchical tension can change according to the tree design, so we expect that a tree can influence the correlation between the objective measure and the subjective ratings.

5   Results and Discussion

M is a weighed linear combination of four terms with weights k_1, k_2, k_3 and k_4 that determine the influence of each corresponding term. We used cross validation to calculate the weights k_i that best fit the scores resulting from the listening test. Among the total of 30 chords for each tree structure that were rated by the listeners, 24 chords were selected to training the 5-fold cross validation, while the rest (6 chords) were applied for the validation part.

---

[2] http://form.jotformeu.com/form/52522142163343.

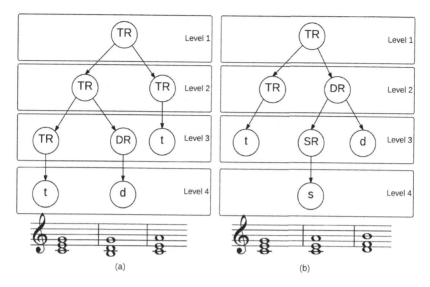

Fig. 4. Representation of the construction of a hierarchical structure for a three-chord progression.

For a particular fold, the set of training sequences might well contain sequences very like the test sequences. To minimize this possibility and increase the reliability of the validation, we calculated the weights and the statistical measures for all the possible combinations of 30 chord progressions of the same tree structure, which makes a total of 4060 possible combinations. The final weights obtained were $k_1 = 4.22$, $k_2 = 2.13$, $k_3 = 2.06$, $k_4 = 3.76$.

Once the weights are incorporated into the measure M for the experiment, we try to validate if the measure M captures the *tonal fitness* of a chord in a progression. In other words, we are evaluating if lower values of M are associated with chords that scored higher in the listening test. Table 1 shows the statistical results (linear regression and error) of the subjective ratings versus the measure M for the chords included in the listening test.

The first column of the table shows the linear regression of the subjective ratings versus the measure M, R^2, along with statistical values calculated from the data, (p-value). The first row contains the results for the chord progressions with the hierarchical structure shown in Fig. 4(a). The second row presents the results for the chord progressions with the tree structure shown in Fig. 4(b). The statistical analysis shows that M captures the tonal fitness of the chords. The p-values are all below the 1% threshold for the null hypothesis, the high R^2 values indicate that the subjective ratings correlate strongly with the objective values.

Likewise, we also include a comparison between our measure M and $L(T_i, P)$ proposed by Lerdahl [8] and $D(T_i, P)$ from a previous work [11]. Both $L(T_i, P)$ and $D(T_i, P)$ are designed as a linear combination of different terms that encode musical factors. $L(T_i, P)$ in the second column considers *consonance*, *tonal tension*, *melodic attraction* or *voice leading* and *distance to previous chord* to calculate the *tonal fitness* of a chord in a different representation space. Thus, the goal of the comparison with $L(T_i, P)$ and $M(T_i, P)$ is to investigate how the representation space can affect to the correlation values between subjective ratings and the evaluation measure. In turn, $D(T_i, P)$ in the third column considers only the *consonance*, the *distance to previous chord* in the progression, the *distance to the key* and the *distance to the tonal function* [11] in the TIS. $D(T_i, P)$ measures the *tonal fitness* of a chord T_i by considering the linear dimension, ignoring hierarchical relationships between the chords. Therefore, we aim to analyze how the hierarchical and linear dimensions encoded in M can influence the subjective ratings against considering only the linear dimension encoded in D, in the same representation space.

We used the results of the same listening test to calculate the optimal weights of each term of L and D via cross validation. The statistical results of L and D are shown in columns two and three of Table 1, respectively, and are also grouped by the tree structure of the chord progressions considered.

Table 1. Comparative table of the statistics R^2 and the p-value for the measures $M(T_i, P)$, Lerdahl's $L(T_i, P)$ and [11] $D(T_i, P)$

		$M(T_i, P)$	$L(T_i, P)$	$D(T_i, P)$
T.1	R^2	0.94 ± 0.10	0.69 ± 0.45	0.90 ± 0.21
	p-value	0.03 ± 0.15	0.39 ± 0.23	0.17 ± 0.18
T.2	R^2	0.94 ± 0.08	0.59 ± 0.43	0.85 ± 0.27
	p-value	0.07 ± 0.13	0.41 ± 0.26	0.17 ± 0.16

The statistical analysis shows that $M(T_i, P)$ captures the *tonal fitness* of the chords better than L and D. The high R^2 values obtained indicate the linear regression of the subjective ratings as a function of the objective values and the p-values are all below the 1% threshold for the null hypothesis, which suggests that the listeners' rates fit the objective measure well. Note that Lerdahl's measure $L(T_i, P)$ also correlates with the subjective ratings of the listening test, but obtaining slightly lower values for R^2 and p-value. Likewise, the R^2 and p-value in $D(T_i, P)$ are correlated, but below the rates obtained for M.

A limitation of the experimental setup is that the listening test was designed to evaluate the chord in a short harmonic context with only two previous chords. Consequently, a three-chord sequence may be perceived by some listeners as a cadential progression, and the listener may expect chords that somehow solve the tension of the previous chords. A longer progression could be needed to reach stronger conclusions about the *tonal fitness* and how the harmonic context and the hierarchical structure can influence the musical perception of one particular chord.

Another limitation is the tree construction of this approach, which does not consider rhythmic features. Rhythm sometimes can influence the hierarchical perception of which chord is above others in the structure, and therefore can also influence the musical perception of the listeners. Taking these factors out of the equation means that some trees could be theoretically equivalent, but in practice, musically perceived in different ways, and consequently, subjectively rated differently. We plan to investigate how the construction of different trees for a particular chord progression can influence the ratings in a future work.

Despite the limitations of this experiment, one of the contributions of this measure is the beginning of a generalization function that can study a long-term chord progression based on both hierarchical and linear structures in different tonalities. The adaptation of Lerdahl's space to the TIS to explain the tonal properties of a chord besides tonal tension and voice leading can be also considered a contribution for the community. The positive results are encouraging to apply this measure to real analysis problems and computational generative systems.

6   Conclusions and Future Work

In this work, we designed a model that captures the *tonal fitness* of one chord in a particular chord progression encoded in the Tonal Interval Space [3]. We use the work proposed by Rohrmeier [16] to represent the hierarchical structure associated with a chord progression. Then we measure the *tonal fitness* of each chord considering factors such as consonance, the distance from the previous chord, the distance to the key, voice leading, and the hierarchical tension of the chord with respect to the rest of the progression. We performed a listening test to evaluate how well chords fit a progression for two different tree structures in different keys. Statistical analysis showed that the subjective ratings correlate strongly with the objective values, validating the objective measure as a proxy for the subjective measure of *tonal fitness*. We also compared the present model with Lerdahl's [8] and a previous proposal [11]. In this preliminary experiment, the measure proposed here was a good indicator of tonal fitness independently of the tree resulting from the hierarchical structure.

Future work will analyze how longer progressions can influence the results obtained with a new listening test, and how different tree structures affect the subjective results. Likewise, the measure can be applied not only to analyze chord progressions from real music samples, but also can be included in a generative system of chord progressions. In a future work, we will investigate how to improve a system to generate chord progressions using an artificial immune system described in a previous work [11] with this new measure to create chord progressions that follow a given pattern, like a tension curve or a given structural tree.

Acknowledgments. Research partially funded by the project "Co-POEM:Platform for the Collaborative Generation of European Popular Music" (ES01-KA201-064933) under the program Erasmus+: KA201-Strategic Partnership; by the project "Experimentation in music in Portuguese culture: History, contexts and practices in the 20th and 21st centuries" (POCI-01-0145-FEDER-031380) co-funded by the European Union through the Operational Program Competitiveness and Internationalization, in its ERDF component, and by national funds, through the Portuguese Foundation for Science and Technology; and by the European Union's Horizon 2020 research and innovation program under the Marie Skłodowska-Curie grant agreement No. 831852 (MORPH).

References

1. Agmon, N.: The Grotthuss mechanism. Chem. Phys. Lett. **244**(5), 456–462 (1995)
2. Amiot, E.: Music Through Fourier Space. Springer, Cham (2016). https://doi.org/10.1007/978-3-319-45581-5
3. Bernardes, G., Cocharro, D., Caetano, M., Guedes, C., Davies, M.E.: A multi-level tonal interval space for modelling pitch relatedness and musical consonance. J. New Music Res. **45**(4), 281–294 (2016)
4. Bernardes, G., Cocharro, D., Guedes, C., Davies, M.E.: Harmony generation driven by a perceptually motivated tonal interval space. Comput. Entertain. (CIE) **14**(2), 6 (2016)
5. Bernstein, L.: The Unanswered Question: Six Talks at Harvard. Harvard University Press, Cambridge (1976)
6. Callender, C., Quinn, I., Tymoczko, D.: Generalized Chord Spaces. Princeton University, Princeton (2006)
7. Herremans, D., Sörensen, K., Martens, D.: Classification and generation of composer-specific music using global feature models and variable neighborhood search. Comput. Music J. **39**(3), 71–91 (2015)
8. Lerdahl, F.: Tonal Pitch Space. Oxford University Press, Oxford (2005)
9. Lerdahl, F., Krumhansl, C.L.: Modeling tonal tension. Music Percept. Interdisc. J. **24**(4), 329–366 (2007)
10. Müllensiefen, D., Wiggins, G.: Polynomial functions as a representation of melodic phrase contour. na (2011)
11. Navarro-Cáceres, M., Caetano, M., Bernardes, G., de Castro, L.N.: ChordAIS: an assistive system for the generation of chord progressions with an artificial immune system. Swarm Evol. Comput. **50**, 100543 (2019)
12. Pachet, F.: The continuator: musical interaction with style. J. New Music Res. **32**(3), 333–341 (2003)
13. Quick, D., Hudak, P.: A Temporal Generative Graph Grammar for Harmonic and Metrical Structure. Michigan Publishing, University of Michigan Library, Ann Arbor (2013)
14. Quinn, S., Watt, R.: The perception of tempo in music. Perception **35**(2), 267–280 (2006)
15. Riemann, H., Haggh, R.H.: History of Music Theory. University of Nebraska, Lincoln (1962)
16. Rohrmeier, M.: Towards a generative syntax of tonal harmony. J. Math. Music **5**(1), 35–53 (2011)
17. Schenker, H., Oster, E.: Free Composition: Volume III of New Musical Theories and Fantasies. Pendragon Press, Hillsdale (1979)
18. Schoenberg, A.: Theory of Harmony. University of California Press, Berkeley (1978)
19. Steedman, M.: The blues and the abstract truth: music and mental models. In: Mental Models in Cognitive Science, pp. 305–318 (1996)
20. Whorley, R.P., Wiggins, G.A., Rhodes, C., Pearce, M.T.: Multiple viewpoint systems: time complexity and the construction of domains for complex musical viewpoints in the harmonization problem. J. New Music Res. **42**(3), 237–266 (2013)
21. Woolhouse, M.: Modelling tonal attraction between adjacent musical elements. J. New Music Res. **38**(4), 357–379 (2009)

Coevolving Artistic Images Using OMNIREP

Moshe Sipper[1,2(✉)], Jason H. Moore[1], and Ryan J. Urbanowicz[1]

[1] Institute for Biomedical Informatics, University of Pennsylvania,
Philadelphia, PA 19104, USA
sipper@gmail.com
[2] Department of Computer Science, Ben-Gurion University,
84105 Beer Sheva, Israel
https://epistasis.org/

Abstract. We have recently developed OMNIREP, a coevolutionary algorithm to discover *both* a representation and an interpreter that solve a particular problem of interest. Herein, we demonstrate that the OMNIREP framework can be successfully applied within the field of evolutionary art. Specifically, we coevolve representations that encode image position, alongside interpreters that transform these positions into one of three pre-defined shapes (chunks, polygons, or circles) of varying size, shape, and color. We showcase a sampling of the unique image variations produced by this approach.

Keywords: Evolutionary algorithms · Evolutionary art · Cooperative coevolution · Interpretation

1 Introduction

Evolutionary art is a branch of evolutionary computation (EC) wherein artwork is generated through an evolutionary algorithm. It is a growing domain, which has boasted a specialized conference over the past few years [4] and many impressive results [8, 26, 27, 36].

In the present study we focus on the evolution of artistic images. To this end, there are generally three major branches of artistic image evolution, differentiated by the standard of 'beauty' applied to drive the fitness of evolving images. The first relies on subjective, interactive feedback from a user [6, 29, 30].

The second approach relies on a target 'inspiration' image to drive fitness. For example, work by [18] used what is essentially a $1 + 1$ evolution strategy—single parent, single child, both competing against each other—to evolve a replica of the Mona Lisa using semi-transparent polygons. This type of evolved art can produce beautiful abstractions of existing pieces of art, or potentially hybrids

This work was supported by National Institutes of Health (USA) grants LM010098, LM012601, AI116794.

© Springer Nature Switzerland AG 2020
J. Romero et al. (Eds.): EvoMUSART 2020, LNCS 12103, pp. 165–178, 2020.
https://doi.org/10.1007/978-3-030-43859-3_12

of existing images. A series of images sampled during the evolutionary process targeting an inspiration image can also offer an artistically appealing output.

The third approach to artistic image evolution incorporates aesthetic measures in the fitness function (light, saturation, hue, symmetry, complexity, entropy, and more) [5,11,14,23].

Returning to the evolutionary methodology, one of the EC practitioner's foremost tasks is to identify a representation—a data structure—and its interpretation, or encoding. These can be viewed, in fact, as two distinct tasks, though they are usually dealt with simultaneously. To wit, one might define the representation as a bitstring and in the same breath go on to state the encoding or interpretation, e.g., "the 120-bit bitstring represents 6 numerical values, each encoded by 20 bits, which are treated as signed floating-point values".

We have recently developed OMNIREP, a coevolutionary algorithm framework to discover *both* a representation and an interpreter that solve a particular problem of interest [32]. We applied OMNIREP successfully to regression and program-evolution tasks. Herein, we demonstrate that OMNIREP can be fruitfully applied within the field of evolutionary art. While the interpreter-representation distinction is perhaps less striking here than with other problems studied by us, we believe both the results and the future possibilities are worthy of presentation, as they demonstrate the efficacy of an alternative and flexible framework for generating evolved art. To the best of our knowledge, this is the first evolutionary art strategy adopting a cooperative (mutualistic) coevolutionary approach, however, some previous work has explored the application of competitive (host-parasite) coevolution to evolving images [11].

In the next section we discuss coevolution and its use as a basis for OMNIREP. Section 3 briefly discusses some previous work. Section 4 presents the OMNIREP algorithm and its application to evolving artful pictures. Results are shown in Sect. 5, followed by concluding remarks in Sect. 6.

2   Coevolution and OMNIREP

In this paper we consider two tasks—discovering a representation and discovering an interpreter—as distinct yet tightly coupled: *A representation is meaningless without an interpretation; an interpretation is useless without a representation.* Our basic idea herein is to employ coevolution to evolve the two simultaneously.

Coevolution refers to the simultaneous evolution of two or more species with coupled fitness [24]. Coevolving species usually compete or cooperate, with a third form of coevolution being commensalism, wherein members of one species gain benefits while those of the other species neither benefit nor are harmed [33] (Fig. 1).

In a competitive coevolutionary algorithm the fitness of an individual is based on direct competition with individuals of other species, which in turn evolve separately in their own populations. Increased fitness of one of the species implies a reduction in the fitness of the other species [15].

A cooperative coevolutionary algorithm involves a number of independently evolving species, which come together to obtain problem solutions. The fitness

(a)   (b)   (c)

Fig. 1. Coevolution: (a) cooperative: Purple-throated carib feeding from and pollinating a flower (credit: Charles J Sharp, https://commons.wikimedia.org/wiki/File:Purple-throated_carib_hummingbird_feeding.jpg); (b) competitive: predator and prey—a leopard killing a bushbuck (credit: NJR ZA, https://commons.wikimedia.org/wiki/File:Leopard_kill_-_KNP_-_001.jpg); (c) commensalistic: Phoretic mites attach themselves to a fly for transport (credit: Alvesgaspar, https://en.wikipedia.org/wiki/File:Fly_June_2008-2.jpg). (Color figure online)

of an individual depends on its ability to collaborate with individuals from other species [7,24,25].

The basic idea of OMNIREP can be stated simply: Rather than specify a specific representation along with a specific interpreter in advance, we shall set up a cooperative coevolutionary algorithm to coevolve the two, with a population of representations coevolving alongside a population of interpreters.[1]

Of importance to note is OMNIREP's not being a specific algorithm but rather an algorithmic framework, which can hopefully be of use in various settings. We believe that the OMNIREP methodology can aid researchers not only in solving *specific* problems but also as an *exploratory* tool when one is seeking out a good representation [32].

3   Previous Work

Generative and Developmental Encoding is a branch of EC concerned with genetic encodings motivated by biology. A structure that repeats multiple times can be represented by a single set of genes that is reused in a genotype-to-phenotype mapping [2,12,13,16,19,20,34,35].

In Gene Expression Programming the individuals in the population are encoded as linear strings of fixed length, which are afterwards expressed as non-linear entities of different sizes and shapes (i.e., simple diagram representations or expression trees) [9].

Though not used extensively, variable-length genomes have been around for quite some time (of course, some representations, such as trees in genetic programming, are inherently variable-length; herein, we simply refer to the literature on "variable-length genomes") [10,21].

---

[1] OMNIREP derives from 'OMNI'—universal, and 'REP'—representation; it also denotes an acronym: Originating MeaNing by coevolving Interpreters and REPresentations.

Grammatical Evolution (GE) was introduced by [28] as a variation on genetic programming. Here, a Backus-Naur Form (BNF) grammar is specified that allows a computer program or model to be constructed by a simple genetic algorithm operating on an array of bits. The GE approach is appealing because only the specification of the grammar needs to be altered for different applications. One might consider subjecting the grammar encoding to evolution in an OMNIREP manner (as done, e.g., by [1]).

Within a memetic computing framework, Iacca et al. [17] proposed, "a bottom-up approach which starts constructing the algorithm from scratch and, most importantly, allows an understanding of functioning and potentials of each search operator composing the algorithm." Caraffini et al. [3] proposed a computational prototype for the automatic design of optimization algorithms, consisting of two phases: a problem analyzer first detects the features of the problem, which are then used to select the operators and their links, thus performing the algorithmic design automatically. Both these works share the desire to tackle basic algorithmic design issues in a (more) automatic manner.

Tangentially related to our work herein is the extensive research on parameters and hyper-parameters in EC, some of which has focused on self-adaptive algorithms, wherein the parameters to be adapted are encoded into the chromosomes and undergo crossover and mutation. The reader is referred to [31] for a comprehensive discussion of this area. Another tangential connection is to "smart" crossover and mutation operators, wherein, interestingly, coevolution has also been applied [37].

4   Evolving Art Using OMNIREP

OMNIREP uses cooperative coevolution with two coevolving populations, one of representations, the other of interpreters. The evolution of each population is identical to a single-population evolutionary algorithm—except where fitness is concerned (Fig. 2). One might argue that the distinction between "representation" and "interpretation" is a malleable one, but this distinction is often put to good use in computer science [32].

We describe below in detail the components and parameters of the OMNIREP system, including: population composition, initialization, selection, crossover, mutation, fitness, elitism, evolutionary rates, and parameters.

Populations. To showcase the application of OMNIREP to evolutionary art, we designed three relatively simple interpreter-representation setups, which produced quite striking results; we refer to them as: chunks, polygons, and circles (Fig. 3). These setups evidence the ease with which OMNIREP can be applied beneficially.

Chunks. An evolving image is composed of linear chunks. A two-dimensional image of dimensions $\{width, height\}$ is treated as a one-dimensional list of pixels of size $width \times height$ (ranging from 10848 pixels to 68816, depending on the particular inspiration image selected for fitness). The representation individual's

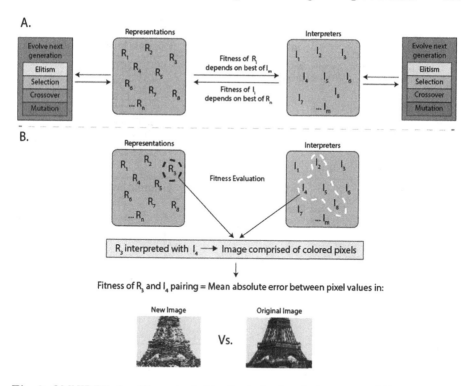

Fig. 2. OMNIREP algorithm adapted to the task of evolutionary art. (A). OMNIREP includes two coevolving populations, one with candidate representations, and the other with candidate interpreters. Each population is evolved using the same fundamental evolutionary algorithm mechanisms (as summarized in the purple boxes). (B) The fitness of a given representation depends on representative interpreters (i.e., the 4 interpreters with the best fitness from the previous generation). In this example R_3's fitness is the average fitness of the four representation-interpreter pairings. The fitness of a given interpreter (e.g., I_3) similarly depends on representative representations (not shown). The fitness of a representation-interpreter pair is computed by combining a representation individual (R) with an interpreter individual (I) to produce the pixel values of an image. Pair fitness is the mean absolute error between the pixels of the new image vs. the inspiration image. (Color figure online)

genome is a list of pixel indexes, $p_i \in \{0, \ldots, width * height - 1\}$, $i = 1, \ldots, 5000$, where p_i is the start of a same-color chunk of pixels. The interpreter individual is a list equal in length to the representation individual, consisting of tuples (b_i, c_i), where b_i is chunk i's length, and c_i is chunk i's color. The image-producing process moves sequentially through the list of chunks, coloring pixels i through $i + b_i - 1$ with color c_i. If a pixel is uncolored by any chunk it is assigned a default base color.

Thus an interpreter individual combines with a representation individual to paint a picture, made up of same-color chunks of length and color indicated by the former and start positions indicated by the latter.

Chunks

Representation	Interpreter	Example Image Output
Pixel Index (e.g., 51)	Length (contiguous pixels) (e.g., 12) Color (of chunk) (e.g., Yellow)	

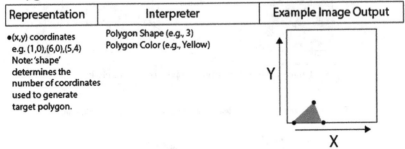

Polygons

Representation	Interpreter	Example Image Output
• (x,y) coordinates e.g. (1,0),(6,0),(5,4) Note: 'shape' determines the number of coordinates used to generate target polygon.	Polygon Shape (e.g., 3) Polygon Color (e.g., Yellow)	

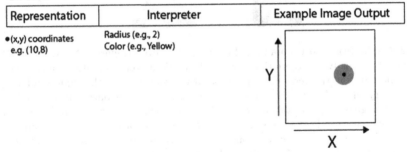

Circles

Representation	Interpreter	Example Image Output
• (x,y) coordinates e.g. (10,8)	Radius (e.g., 2) Color (e.g., Yellow)	

Fig. 3. Overview and examples of the three image-mapping strategies employed by OMNIREP in experimental evolutionary runs.

Polygons. An evolving image is composed of polygons. The representation individual's genome is a list of 600 polygon coordinates, $[x_i, y_i]$, $i \in \{1, \ldots, 600\}$. The number of polygons was set to 50 and the maximum number of sides to 12, hence 600. The interpreter individual is a list of length 50, representing 50 polygons, consisting of tuples (s_i, c_i), where s_i is polygon i's shape, i.e., number of sides, and c_i is polygon i's color.

An interpreter individual combines with a representation individual to paint a picture, made up of 50 polygons whose shape and color are indicated by the former and coordinates by the latter. The image-producing process moves sequentially through the interpreter and representation genomes, picking the appropriate number of coordinates determined by s_i, and coloring the resultant shape

Fig. 4. Inspirational images. (Color figure online)

with color c_i. Coordinates "left over" in the representation genome in the end are unused.

Circles. An evolving image is composed of circles. The representation individual's genome is a list of 50 circle centers, $[x_i, y_i]$, $i \in \{1, \ldots, 50\}$. The interpreter individual is a list equal in length to the representation individual, consisting of tuples (c_i, r_i), where c_i is circle i's color and r_i is its radius.

An interpreter individual combines with a representation individual to paint a picture, made up of 50 circles whose colors and radii are indicated by the former and centers by the latter.

For chunks, polygons, and circles, if there is any overlap in these shapes forming the image, the most recently placed shape opaquely covers the overlapped shape.

Initialization. For every coevolutionary run: both populations are initialized to random values in the appropriate range, depending on the particulars of the interpreter-representation setup delineated above (chunks, polygons, or circles). An inspiration image is chosen per evolutionary run as one of the 8 shown in Fig. 4. The inspiration images were converted to 4 colors from their originals using Python's `Image.ADAPTIVE` palette (in the `PIL` package). Note that the 4 colors differ between images.

Selection. Tournament selection with tournament size 4, i.e., choose 4 individuals at random from the population and return the individual with the best fitness as the selected one.

Crossover. Single-point crossover—select a random crossover point and swap two parent genomes beyond this point to create two offspring—is employed every generation.

Mutation. Mutation is done with probability 0.3 (per individual) by selecting a random gene and replacing it with a new random value in the appropriate range.

Fitness. To compute fitness the two coevolving populations cooperate. Specifically, to compute the fitness of a single individual in one population, we use *representatives* from the other population [24]. The representatives (also called cooperators) are selected via a greedy strategy as the 4 fittest individuals from the previous generation. When evaluating the fitness of a particular representation individual, we combine it 4 times with the top 4 interpreter individuals, compute 4 fitness values, and use the average fitness over these 4 evaluations as the final fitness value of the representation individual. In a similar manner we use the average of 4 representatives from the representations population when computing the fitness of an interpreter individual. (Other possibilities include using the best fitness of the 4, the worst fitness, and selecting a different mix of representatives, e.g., best, median, and worst.)

A representation individual and an interpreter individual are combined as described above per interpreter-representation setup (chunks, polygons, or circles). A single fitness value then equals the mean absolute error with respect to the known inspirational pixels (Fig. 2).

Note that our objective was not to reproduce the original image precisely—which would be uninteresting. Rather, the selected image serves as inspiration, setting an evolutionary direction.

Elitism. The 2 individuals with the highest fitness in a generation are copied ("cloned") into the next generation unchanged.

Evolutionary rates differ between the two populations, with the interpreters population evolving more slowly, specifically, every 3 generations.

Parameters. Table 1 provides a summary of parameters discussed throughout the paper.[2]

5   Results

We performed multiple evolutionary runs for each of the 8 target images. Figures 5, 6 and 7 present select examples of the images evolved with OMNIREP. Each figure gives sample outputs from each of the three underlying setups utilized (i.e., chunks, circles, or polygons). Each set of 4 images (derived from a specific inspiration image) serves two purposes: (1) demonstrating the evolutionary trajectories of evolving art, namely, images in intermediate generations of an evolutionary run, and (2) presenting the images as part of a 'panel', i.e., a collection of images in series, meant to be viewed as a single artistic piece. Indeed, subjectively, we feel that part of the appeal of the images evolved by OMNIREP is the progression of an image from a state of abstraction to one closer to the inspirational piece. Of course, individual images generated with OMNIREP can also be selected as the output artistic piece.

---

[2] Some parameters may seem arbitrary but our recent findings provide some justification for this [31].

Table 1. Evolutionary parameters. Shown first are common parameters, followed by experiment-specific ones.

Description	Value
Common	
Number of images	8
Size of representations population	20
Size of interpreters population	10
Type of selection	Tournament
Tournament size	4
Type of crossover	Single-point
Probability of mutation (representations)	0.3
Probability of mutation (interpreters)	0.3
Evolve interpreters population every	3 generations
Number of representatives used for fitness	4
Number of top individuals copied (elitism)	2
Number of colors	4
Chunks	
Number of generations	20000
Size of representation individual	5000
Size of interpreter individual	5000 (chunks)
Minimum chunk size	1 pixel
Maximum chunk size	10 pixels
Polygons	
Number of generations	50000
Size of representation individual	600
Size of interpreter individual	50 (polygons)
Minimum polygon sides	3
Maximum polygon sides	12
Circles	
Number of generations	50000
Size of representation individual	50
Size of interpreter individual	50 (circles)
Minimum radius	3
Maximum radius	50

Beyond generating static images, we have also explored converting the set of evolving images into appealing animated GIFs (see samples at https://github.com/EpistasisLab/OMNIREP).

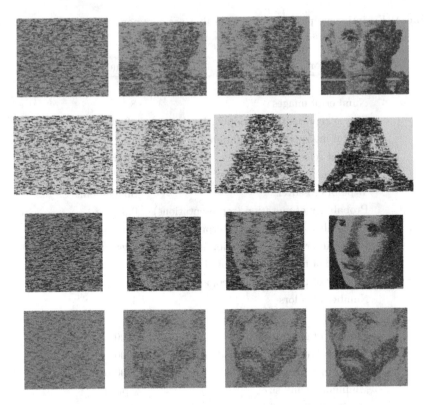

Fig. 5. Evolutionary trajectories of evolving images (chunks). Each row represents a single run.

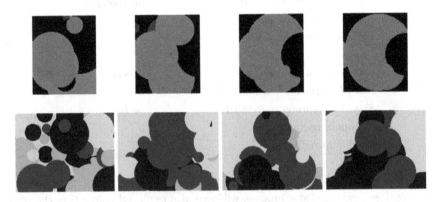

Fig. 6. Evolutionary trajectories of evolving images (circles). Each row represents a single run.

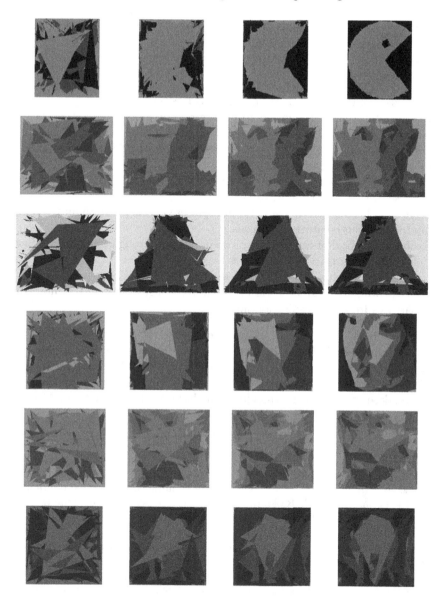

Fig. 7. Evolutionary trajectories of evolving images (polygons). Each row represents a single run.

6   Concluding Remarks

We adapted and applied OMNIREP, which coevolves representations and interpreters simultaneously, to evolutionary art. This work demonstrates the potential of a mutualistic coevolutionary system, as well as a strategy that separates representations from interpreters, towards the task of evolved artistic images. This framework can be flexibly extended in the future to generate an even greater diversity of novel and intriguing imagery.

Some of the immediate future directions we expect would be valuable to explore include:

1. Expansion of the color palette beyond 4 colors per image.
2. Giving interpreters the option to choose one or more shapes (i.e., chunks, polygons, or circles) to incorporate into a given evolved image.
3. Expand the shape options to include other forms or orientations (e.g., ellipsoids, squares, rectangles, etc.).
4. Promote overlapping shapes in the image and assign color mixes to the image in these overlapping regions.
5. More complex interpreters, e.g., an interpreter genetic programming tree that interprets a representation of pixels.
6. Evolving a hybrid of multiple inspiration images through integration of multiobjective optimization.
7. Incorporate novelty [22] into the coevolutionary algorithm (as proposed in [33]) to promote a greater variety of novel images that create interesting departures from the inspiration image.

We perceive OMNIREP not as a particular algorithm but rather as a meta-algorithm, which might hopefully be suitable for other settings, beyond the artful one described herein. Essentially, any scenario where some form of representation may be interpreted in several ways, or where the representation and interpreter can be rendered "fluid" rather than fixed, might be a candidate for an OMNIREP approach.

References

1. Azad, R.M.A., Ryan, C.: An examination of simultaneous evolution of grammars and solutions. In: Yu, T., Riolo, R., Worzel, B. (eds.) Genetic Programming Theory and Practice III. GPEM, vol. 9, pp. 141–158. Springer, US, Boston, MA (2006). https://doi.org/10.1007/0-387-28111-8_10
2. Bentley, P., Kumar, S.: Three ways to grow designs: a comparison of embryogenies for an evolutionary design problem. In: Proceedings of the 1st Annual Conference on Genetic and Evolutionary Computation, GECCO 1999, vol. 1, pp. 35–43. Morgan Kaufmann Publishers Inc., San Francisco (1999). http://dl.acm.org/citation.cfm?id=2933923.2933928
3. Caraffini, F., Neri, F., Picinali, L.: An analysis on separability for memetic computing automatic design. Inf. Sci. **265**, 1–22 (2014)
4. Correia, J., Ciesielski, V., Liapis, A.: Proceedings of Computational Intelligence in Music, Sound, Art and Design: 6th International Conference. Springer, Berlin (2017). https://doi.org/10.1007/978-3-319-55750-2
5. Datta, R., Joshi, D., Li, J., Wang, J.Z.: Studying aesthetics in photographic images using a computational approach. In: Leonardis, A., Bischof, H., Pinz, A. (eds.) ECCV 2006. LNCS, vol. 3953, pp. 288–301. Springer, Heidelberg (2006). https://doi.org/10.1007/11744078_23

6. Dawkins, R.: The Blind Watchmaker: Why the Evidence of Evolution Reveals a Universe Without Design. WW Norton & Company, New York (1996)
7. Dick, G., Yao, X.: Model representation and cooperative coevolution for finite-state machine evolution. In: 2014 IEEE Congress on Evolutionary Computation (CEC), pp. 2700–2707. IEEE, Piscataway (2014)
8. DiPaola, S., Gabora, L.: Incorporating characteristics of human creativity into an evolutionary art algorithm. Genet. Program Evolvable Mach. 10(2), 97–110 (2009). https://doi.org/10.1007/s10710-008-9074-x
9. Ferreira, C.: Gene expression programming: a new adaptive algorithm for solving problems. Complex Syst. 13(2), 87–129 (2001)
10. Goldberg, D.E., Korb, B., Deb, K.: Messy genetic algorithms: motivation, analysis, and first results. Complex Syst. 3, 493–530 (1989)
11. Greenfield, G.R.: Simulated aesthetics and evolving artworks: a coevolutionary approach. Leonardo 35(3), 283–289 (2002)
12. Gruau, F., Whitley, D., Pyeatt, L.: A comparison between cellular encoding and direct encoding for genetic neural networks. In: Proceedings of the 1st Annual Conference on Genetic Programming, pp. 81–89. MIT Press, Cambridge (1996). http://dl.acm.org/citation.cfm?id=1595536.1595547
13. Hart, W.E., Kammeyer, T.E., Belew, R.K.: The role of development in genetic algorithms. In: Whitley, L.D., Vose, M.D. (eds.) Foundations of Genetic Algorithms, vol. 3, pp. 315–332. Elsevier (1995). https://doi.org/10.1016/B978-1-55860-356-1.50019-4. http://www.sciencedirect.com/science/article/pii/B9781558603561500194
14. den Heijer, E., Eiben, A.E.: Comparing aesthetic measures for evolutionary art. In: Di Chio, C., et al. (eds.) EvoApplications 2010. LNCS, vol. 6025, pp. 311–320. Springer, Heidelberg (2010). https://doi.org/10.1007/978-3-642-12242-2_32
15. Hillis, W.: Co-evolving parasites improve simulated evolution as an optimization procedure. Physica D: Nonlinear Phenomena 42(1), 228–234 (1990)
16. Hornby, G.S., Pollack, J.B.: Creating high-level components with a generative representation for body-brain evolution. Artif. Life 8(3), 223–246 (2002)
17. Iacca, G., Neri, F., Mininno, E., Ong, Y.S., Lim, M.H.: Ockham's razor in memetic computing: three stage optimal memetic exploration. Inf. Sci. 188, 17–43 (2012)
18. Johansson, R.: Genetic programming: evolution of Mona Lisa (2008). www.rogerjohansson.blog/2008/12/07/genetic-programming-evolution-of-mona-lisa/. Accessed 23 Apr 2018
19. Koza, J.R.: Genetic Programming IV: Routine Human-Competitive Machine Intelligence. Kluwer Academic Publishers, Norwell (2003)
20. Koza, J.R., Andre, D., Bennett, F.H., Keane, M.A.: Genetic Programming III: Darwinian Invention & Problem Solving, 1st edn. Morgan Kaufmann Publishers Inc., San Francisco (1999)
21. Lee, C.Y., Antonsson, E.K.: Variable length genomes for evolutionary algorithms. In: Proceedings of the Genetic and Evolutionary Computation Conference. Morgan Kaufmann (2000)
22. Lehman, J., Stanley, K.O.: Exploiting open-endedness to solve problems through the search for novelty. In: In Proceedings of the Eleventh International Conference on Artificial Life (ALIFE). MIT Press (2008)
23. Machado, P., Romero, J., Manaris, B.: Experiments in computational aesthetics. In: Romero, J., Machado, P. (eds.) The Art of Artificial Evolution. NCS, pp. 381–415. Springer, Heidelberg (2008). https://doi.org/10.1007/978-3-540-72877-1_18
24. Pena-Reyes, C.A., Sipper, M.: Fuzzy CoCo: a cooperative-coevolutionary approach to fuzzy modeling. IEEE Trans. Fuzzy Syst. 9(5), 727–737 (2001)

25. Potter, M.A., De Jong, K.A.: Cooperative coevolution: an architecture for evolving coadapted subcomponents. Evol. Comput. **8**(1), 1–29 (2000)

26. Romero, J., Machado, P. (eds.): The Art of Artificial Evolution: A Handbook on Evolutionary Art and Music. Natural Computing Series. Springer, Heidelberg (2008). https://doi.org/10.1007/978-3-540-72877-1

27. Romero, J., McDermott, J.: EvoMUSART 2014: third international conference on evolutionary and biologically inspired music, sound, art and design. Leonardo **49**(3), 245–245 (2016)

28. Ryan, C., Collins, J.J., Neill, M.O.: Grammatical evolution: evolving programs for an arbitrary language. In: Banzhaf, W., Poli, R., Schoenauer, M., Fogarty, T.C. (eds.) EuroGP 1998. LNCS, vol. 1391, pp. 83–96. Springer, Heidelberg (1998). https://doi.org/10.1007/BFb0055930

29. Secretan, J., Beato, N., D Ambrosio, D.B., Rodriguez, A., Campbell, A., Stanley, K.O.: Picbreeder: evolving pictures collaboratively online. In: Proceedings of the SIGCHI Conference on Human Factors in Computing Systems, pp. 1759–1768. ACM (2008)

30. Sims, K.: Artificial evolution for computer graphics, vol. 25. ACM (1991)

31. Sipper, M., Fu, W., Ahuja, K., Moore, J.H.: Investigating the parameter space of evolutionary algorithms. BioData Min. **11**(2), 1–14 (2018). https://doi.org/10.1186/s13040-018-0164-x

32. Sipper, M., Moore, J.H.: OMNIREP: originating meaning by coevolving encodings and representations. Memetic Comput. **11**(3), 251–261 (2019). https://doi.org/10.1007/s12293-019-00285-2

33. Sipper, M., Moore, J.H., Urbanowicz, R.J.: Solution and fitness evolution (SAFE): coevolving solutions and their objective functions. In: Sekanina, L., Hu, T., Lourenço, N., Richter, H., García-Sánchez, P. (eds.) EuroGP 2019. LNCS, vol. 11451, pp. 146–161. Springer, Cham (2019). https://doi.org/10.1007/978-3-030-16670-0_10

34. Stanley, K.O., D'Ambrosio, D.B., Gauci, J.: A hypercube-based encoding for evolving large-scale neural networks. Artif. Life **15**(2), 185–212 (2009)

35. Stanley, K.O., Miikkulainen, R.: A taxonomy for artificial embryogeny. Artif. Life **9**(2), 93–130 (2003)

36. Wikipedia: Evolutionary art (2018). https://en.wikipedia.org/wiki/Evolutionary_art

37. Zaritsky, A., Sipper, M.: The preservation of favored building blocks in the struggle for fitness: the puzzle algorithm. IEEE Trans. Evol. Comput. **8**(5), 443–455 (2004)

Sound Cells in Genetic Improvisation: An Evolutionary Model for Improvised Music

Sebastian Trump[(✉)]

Nuremberg University of Music, Nuremberg, Germany
sebastian.trump@hfm-nuernberg.de

Abstract. Musical improvisation and biological evolution are similarly based on the principles of unpredictability and adaptivity. Within this framework, this research project examines whether and how structures of evolutionary developmental logic can be detected and described in free improvisation. The underlying concept of improvisation is participative in nature and, in this light, contains similar generative strategies as there are in evolutionary processes. Further implications of the theory of evolution for cultural development in the concept of memetics and the form of genetic algorithms build an interdisciplinary network of different theories and methodologies, from which the proposed model of genetic improvisation emerges.

Keywords: Evolutionary algorithm · Musical Improvisation · Computational analysis

1 Introduction

Looking at the intersections of music and evolutionary theories, the anthropological perspective is the first to come in mind [33,35]. The evolutionary view of musical development emphasizes their continuous and overarching character: "Perhaps most importantly for ethnomusicologists, metaphors of [...] evolution turn attention away from the agency of individuals." [30]. This perspective draws a direct link to the concepts of computational creativity as a reverse-engineered use of evolutionary algorithms in musical improvisation. Wallin, Merker, and Brown [35], therefore, call for expanding the methodological repertoire of evolutionary musicology by approximating traditional notation-based and computer-aided empirical analysis approaches. This could even help to provide new foundations for the open problem of evolutionary art theories [25].

The objective of the study is to detect and describe structures of evolutionary developmental logic in this process. Biologist Dawkins [9] coined the term "memetics" as a cultural replicator equivalent to genes in biological genetics. Developed into a philosophy of mind by Dennett [11], memetics are relevant for the present investigation and will be tested to see if it can be useful as the philosophical basis of a model of genetic improvisation.

J. Romero et al. (Eds.): EvoMUSART 2020, LNCS 12103, pp. 179–193, 2020.
https://doi.org/10.1007/978-3-030-43859-3_13

In mathematics, generative rules can produce an infinite amount of concrete numbers according to a single abstract scheme. Such formulas relate to numbers like grammar to language [3]. Linguist Chomsky [6] developed the concept of "Transformational-Generative Grammar", a syntax theory of language that generates an infinite set of sentences from a finite set of grammatical rules. While there are several approaches in the field of evolutionary music generation [1,26], this study aims to apply evolutionary theory to the analysis of musical improvisation.

For the analysis of jazz improvisations, the way of transcribing into musical notation is usually chosen and then studied like composed music using conventional methods. Frieler, Lothwesen, and Schütz [15] criticize the narrowed scope in this approach, which barely captures the creative process, and therefore alternatively suggest a data-driven approach [17] by "idea flow"-analysis. Improvisation thereby appears as a chain of musical ideas that can be classified according to various categories and form the musical surface. This "midlevel"-approach [16] seems to be promising, but continues to build on transcribed music texts and in this form can not be transferred to free improvisation without loss of essential components such as timbre variation. Lartillot and Ayari [22], similar to the "Generative Theory of Tonal Music" [23], follow the path of structural modelling [21] of improvisations during listening. A similar modelling approach translates pitch curves into a complex mathematical network [13] and presents the melodic evolution as a more or less dense graph of linked nodes representing single tones. This network-theory based approach also seems to be useful for the methodology developed here. Ferretti's study, however, focusses on individual soloist-oriented improvisation and thus largely neglects aspects of interaction.

Pressing [29] reconstructs a single improvisation as a linear structure in his model and thus offers another possible starting point for the modelling of an intra-improvisational structure, however without explicitly describing interactive and tonal aspects. Complementary to this, Borgo suggests an analysis method based on *corellograms*, which visualize sonic complexity in the course of an improvisation [5].

Building on these analytical approaches, the integration of evolutionary concepts opens up even more perspectives. In addition to the macroscopic view, which is rooted in anthropological theories on the function and genesis of music, it is essential to consider improvisation with microscopic focal length. This makes the genesis and further development of individual *sound cells* comprehensible and at the same time accessible to technical and hermeneutic analysis. The central question for this investigation remains whether the description of free improvisation as an evolutionary developmental logic in a genetic model is helpful, to gain new insights into improvisation in particular, but also into creative interaction processes in general.

2   The Model of Genetic Improvisation

In the model of genetic improvisation presented here, the meme appears in the special form of the *sound cell* as the individual in the evolutionary process. The sound cell is the improvisational counterpart to a composed musical phrase and combines in analogy the realisation instructions of a score as a genotype with the sonic realization as a phenotype [34]. This results in a methodological problem: With the phenomenon of sound as the starting point of analysis, one can only speculate about its underlying mental, here genetic, disposition. This is done by the equally fuzzy method of cluster analysis. However, aspects of sound production through musical instruments and their interactions are not explicitly modelled here as well as those of other external factors such as the influence of the audience.

With this understanding, the sound cell appears as a material group, i.e. a unit of structurally or sonically related material within a genetic possibility space. Thus, sound cells act as "micro musical instruments" with a specific sound potential and with their ontogenesis fixate one actually played phenotype. Their life span is an indicator of the degree of adaptation to the musical context and selection pressure, either by co-sounding sound cells of other improvisers or – in non-free, stylistically bound improvisation – created by the formal and harmonic structure. Together with the reproduction rate as a measure of cultural attractiveness the *fitness* of individual sound cells can be determined.

Like improvisation and evolution, this model can also be described in two opposing directions – on the one hand as a product or fossil, which becomes accessible to analysis and systematization, and on the other hand as a process or algorithm that unfolds its generative potential in the future. The course of a musical improvisation appears retrospectively as an evolutionary development of a population of sound cells, which branch out in a phylogenetic tree on several *lineages*. For a general understanding of this model, it seems helpful to first take the generative perspective:

An improvisation is created by successive sound cells. Each sound cell exists only for the duration of its sound and contains a specific musical potential. Sound cells form as uniparental clones and are transformed by a mental selection process [20] (*creative selection*) between different mutations until the most appropriate sound cell from the descendant generations of all lineages is realized. The actual selection process remains hidden. Predetermined musical structures act as further *selection factors*. Several lineages can arise in parallel and also produce offspring.

If the improvisation takes place with several players, their sound cells develop as a common population, so that lineages can change between improvisers and enable fast interaction and synchronization on different musical levels. The influence of these sound cells as *environmental induction* on the other players simultaneously creates another level of selection from the point of view of the overlapping lineages, but here it is transparent since it takes into account only realized sound cells[1].

For a comprehensive presentation of the model, complementary to the generative-algorithmic description, the retrospective-analytical perspective is necessary, which, like a decrypted archaeological puzzle, completes the overall picture of genetics of sound cells and thus of genetic improvisation as a whole:

(1) A sound cell consists of one or more sounds as a homogeneous unit[2].
(2) The sonic shape of a sound cell is described as its *phenotype* according to fixed acoustic parameters.
(3) The grouped multi-dimensional feature space of the phenotype cumulates at certain points (*clusters*), which form several *alleles* of a gene per cluster group. A gene bundles a musical character trait as a combination of several phenotypic features. This results exclusively in *polymorphic* expressions of genes.
(4) The totality of all appearing alleles in the associated genes forms the *genome* of the entire sound cell population.
(5) The *genotype* of a sound cell encodes all specific alleles.
(6) From the genetic similarity of the sound cells their *phylogenetic structure* can be reconstructed as a tree of several lineages.

In the function of the genotype finally, the analytical and the generative perspective are united. While for a genotype infinite sonic variants within the framework are possible, each phenotype can be assigned to one distinct genotype. As a deep structural design plan of a sound cell, it is also the link in their evolutionary development.

2.1   Computational Analysis

The following analysis aims to apply the proposed model of genetic improvisation as an empirical method to a corpus of improvisations recorded specifically

---

[1] Memetics has developed the notion of "co-adaptive memplexes" for this mode of mutual selection pressure, which can be applied to the interaction of the sound cell lineages. Dawkins writes: "memes, like genes, are selected against the background of other memes in the meme pool. The result is that of mutually compatible memes – coadapted meme complexes or memeplexes – are found cohabiting in individual brains." [2].

[2] The question of a suitable segmentation of the sound material for an analytical access proves to be very complex in detail. Dawkins already hints at the problem of blurring in the segmentation of a meme: "So far I have talked about memes as though it is obvious what a single unit-meme consited of. But of course it is far from obvious. I have said it is a meme, but what about a symphony: how many memes is that? Is each movement one meme, each recognizable phrase of melody, each bar, each chord, or what?" [9]. The model of genetic improvisation follows Jan's considerations on composed music, according to which small units provide easier ways to connect to a "intact imitation" [19].

for this purpose, and then to evaluate the computed results. For the data acquisition, a laboratory situation was created, in which two musicians were spatially isolated from each other and improvised together only through microphones and headphones. This concept follows the idea that any communication and interaction would have to be coded in sound and thus can be tracked in an acoustic analysis.

For the improvisations examined here, the recording setting was transformed into a concert performance while retaining the spatial separation. Thus, the laboratory-specific artificial recording situation could merge with the stimulating atmosphere of a performance situation and create the most authentic possible framework for the emergence of free improvisation. In the audience room, an octophonic loudspeaker circle made improvised duets between the eight participating musicians audible, who were repeatedly combined into pairs randomly by an algorithm. All recordings sum up to a corpus of about 30 h of sound material in 246 improvisations[3].

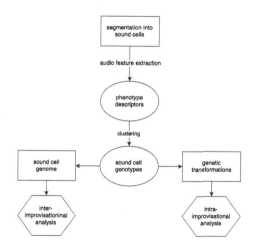

Fig. 1. Flowchart of the analysis process

The analysis of the acquired recordings is done by the custom software package *GenImpro*[4] in several steps (see Fig. 1), first the preparation and segmentation, then the description of the phenotype of the sound cells, the determination of the common genome and their connection as sound cell phylogenetics.

---

[3] For a short documentation of the recording setting see https://vimeo.com/150357914.

[4] https://github.com/bastustrump/genimpro.

2.2   Sound Cell Phenotype

A phenotype describes the totality of all the external characteristics of an individual. For sound cells, the phenotypic description should be determined by the characteristics of the sounds that can be generated by them.

The starting point for the phenotype of a sound cell is the audio features of the *Essentia* library [4]. These often, but not always, correlate directly to an auditory quality of the underlying sound. They are roughly divided into the musical categories of dynamics, rhythm, harmony, melody and timbre. Especially in the latter group of characteristics of the timbre, however, it is apparent that the application areas intended for their development, such as instrument categorization or genre classification, can not reproduce variations that arise within expressive improvising. Here, the data values per improvisation and player are normalized to achieve a higher level of detail. Nevertheless, the attempt to quantify sonic properties always remains an approximation – especially concerning timbre, which as a "complex set of auditory attributes" [24] is strongly linked to subjective perception. Irrelevant to the phenotypic characterization is the duration of a sound cell. This aspect is relevant as a component of the *sound cell fitness*, which, moreover, can only be evaluated in the larger context of sound cell evolution.

Following the goal of the model presented here, to show structures through relative references in improvisational development and thus to achieve their comparability across different idioms, all categories contain attributes that follow the principle of relativization, but not abstraction if possible. The psychoacoustic phenomenon of the "roughness" of a sound and its influence on the perception of musical structure investigated by McAdams [24] is used here in the broader sense as a degree of variation or fluctuation.

The computational approach to the characteristics has, in addition to the representation of sound as a perceptual phenomenon, another function. The technical description of the sounds results in quantitative comparability, for which the assignment of the phenotype characteristics to certain perceptual sound properties is not a necessary prerequisite. Moreover, the individual parameters are not to be considered in isolation, but on the contrary overlap in the properties they describe. This often results in phenotype-immanent correlations that have no artistic-creative origin. Only in the relationship of several sound cells to each other, genetic development is emerging.

2.3   Sound Cell Genome

The design of the sound cell genome follows the principle of combining characteristic properties as specific combinations of several phenotype characteristics. The grouping of features is unique for each gene and only relevant for the genome if there are several significant accumulations in the feature space and thus distinguishable alleles of the gene. The statistical method of cluster analysis is used to calculate the polymorphic variants. The combinations are composed of three groups:

1. All one-dimensional phenotype components are combined in all possible pairs to allow every correlation to be coded into the genotype.
2. The two multi-dimensional phenotype features, pitch-class profile and relative interval structure, each form a separate group.
3. The items from the first step are individually linked to the two pre-grouped combinations from step 2.

The resulting structure consists of 298 genes with a total of 5,800 different alleles (see Fig. 2). As in the phenotype, redundancies in the genome also result from overlapping representations, which, however, reproduce their musical nuances and thus build up a sufficiently expressive capacity of the genome as a whole and for each genotype. With the genome structure formed here, a theoretical number of $1.6 * 10^{353}$ of different genotypes could be mapped.

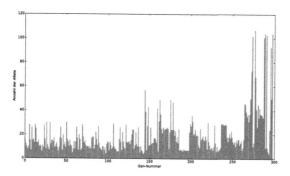

Fig. 2. Number of alleles in the sound cell genome

2.4 Phylogenetics

Phylogenetic methods reconstruct the phylogenetic evolution of species or individual organisms based on their genetic similarity [14,18]. In a *phylogenetic tree*, relationships between the examined taxa (e.g. species or single genes) are shown[5].

For the model of genetic improvisation, established methods of bioinformatics are applied[6] to analyse the evolutionary relationships of a sound cell. The totality of all sound cells forms a gene pool that can be subdivided to consider specific aspects of their interaction. The method of *neighbour-joining* [31] generates an unrooted phylogenetic tree of genotypes based on a similarity matrix of the gene pool through hierarchical clustering. The algorithm for deriving the tree

---

[5] The most common methods are (1.) UPGMA *unweighted pair-group method using arithmetic averages* or *neighbor-joining* [31] for calculating an initial tree using a distance matrix [12], (2.) the *parsimony principle*, in which several possible trees are named after the number of trees and (3.) *Maximum Likelihood*, which selects the most probable of random trees.

[6] In GenImpro the *Biopython* package is integrated [7].

assumes that all sound cells are located as leaves at the end of the branches and intermediate stages exist only in the form of internal nodes. In fact, however, the temporal succession is known for use in intra-improvisational analysis, so only the coarse subdivision into lineages is extracted and integrated into the real order. This results in a topology of partially linked, partly unlinked lineages.

Another level of genetic analysis is revealed when multiple sound cells and their individual genotypes are combined into gene pools. Thus, the average genotype is a condensed image of the included sound cells, which makes inter-improvisational connections visible on a larger scale. The highly complex data generated from the analyses in the previous sections, on the one hand, has little significance in isolated view, but on the other hand is difficult to grasp in inter-action. To address this, an integrated browser-based visualization is integrated into *GenImpro*, which serves as an interactive research environment and visual-isation of partial results. The selection of a certain recording can be done via networks of differently grouped sound cells, which are summarized according to the desired perspective according to their origin or their affiliation to a certain improvisation. The resulting gene pools as a similarity matrix give an initial overview of evolutionary relationships.

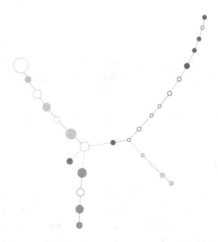

Fig. 3. Representation of an improvisation as an expanded phylogenetic tree (Color figure online)

Colour-coded by lineage, the sound cells are organized as circles in the order of their appearance along the branches. The personal origin of the sound cells can be distinguished by a filled or rimmed shape whose size depends on the lifetime. The length of the connecting lines refers to the genetic similarity (see Fig. 3).

To reproduce the exact temporal course of the improvisation, one can switch to a second mode, which breaks up the tree structure in favour of a linear time axis and the separation according to player (see Fig. 4). The interaction between

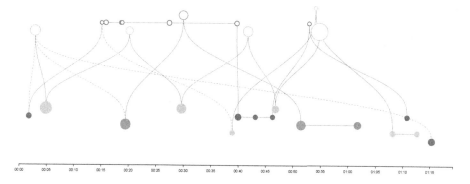

Fig. 4. Representation of the sound cells in chronological order

the players as well as the sound cell lineages become clearly visible through connections between the upper and lower half. Besides, the emergence of this structure as animation can be traced simultaneously while listening.

3   Discussion of the Results

GenImpro opens up the possibility to approach the corpus of improvisations from different perspectives and with varying degrees of detail. Although these perspectives always seem to be interlinked, they should be viewed as differentiated approaches. In addition to the genetic exploration of individual improvisations[7], systematic access to the entire corpus becomes possible. The data of the model offers a variety of partly bio-genetically inspired, partly improvisation specific starting points.

Population genetics has constructed the unit of the gene pool for the analysis of macro-genetic relationships in which local and global genetic phenomena can be linked. This compact and consistent description of each improvisation creates a quantitative level of comparison for free improvisations.

Each improvisation appears in the tree (see Fig. 5) as a combination of two circles whose colours refer to the musicians involved. The surrounding background of this double symbol indicates with its grey value the average interactivity of all contained lineages[8].

The gene pool of an improvisation describes as its "genetic fingerprint" the exact distribution of alleles within the considered group and thus allows the combination of otherwise difficult to grasp structures and musical characteristics. Furthermore, improvisation groups with common genetic characteristics become visible in the branches of the tree. To investigate possible distinction criteria, the

---

[7] Examples of an hermeneutic interpretation of intra-improvisational results are presented on http://www.genimpro.net/intra-improvisational.

[8] An interactive version of this visualisation can be found on http://www.genimpro.net.

Fig. 5. Phylogenetic tree of all improvisation gene pools (Color figure online)

tree was subdivided into nine groups (highlighted in grey colour) and analysed for identical genes.

A lineage can arise from sound cells of several improvisation participants and has a specific measure of their *interactivity* due to the frequency of the change between players. The gene transfer resulting from this exchange potential between the interacting sound cells does not have to lead to a musical imitation in the sense of clear external similarity but is usually only partially effective as a *genetic imitation*. Overall, however, lineages with intensive interaction also show high genetic homogeneity.

In addition to lineage interactivity, there are also other factors of selection for the individual sound cell, which can be summarized as their *sound cell fitness*. It is composed of (1.) the length of the sound cell, (2.) the number of descendants as the remaining length of the associated lineage, and (3.) the interactivity of the lineage.

Another starting point for systematic access to the data of the model is the recombination of sound cells into categorical gene pools. Due to the manageable group size and the associated comparability, individual gene pools were formed from the sound cells of the 15 players and examined with the described analysis modes.

Particularly interesting in the inspection of the player-specific values (see Fig. 6) is that although genetic imitation increases the fitness of the sound cells, it does not equally increase their interaction potential. On the contrary, an opposite

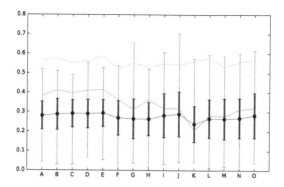

Fig. 6. Average fitness with ranges of values and standard deviations of the sound cells (blue). Genetic imitation degree (green). Interactivity (red). (A: Snaredrum, B: Snaredrum, C: Snaredrum, D: Snaredrum, E: Snaredrum, Q: recorder, G: recorder, H: recorder, I: violin, J: double bass, K: sounds, L: percussion, M: trombone, N: cello, O: soprano saxophone/flute) (Color figure online)

effect seems to be effective, which rewards a similar style of play in the short term, but in the course of the improvisation it restricts itself through less interaction.

Using sound cell fitness and lineage interactivity, two inter-improvisational benchmarks can be derived from the model, adding insightful information to the different aggregation levels – from the entire corpus to player typologies to individual improvisation. Regarding the research question of this investigation, these macro-analytical aspects are a significant gain for the description of improvisation as an evolutionary process.

A gain in knowledge in the consideration of a single improvisation is given especially by the fact that analogies and relationships can be established based on the genetic disposition of individual sound cells. This can already be seen in the first abstraction stage of lineages, which, following the logic of analysis, group together according to a similar genotype. Despite many recognition errors in the computation of phenotypes, the chosen step of abstraction to single genes proves to be sufficiently robust to show interesting structures.

Although a hermeneutic approach remains highly speculative, it allows specific characteristics to be attributed to individual genes. The cluster groups of the phenotypic feature space seem to be losing relevance in the overview and instead to represent an overall musical quality in larger gene combinations. To understand the sound cells as individuals struggling to survive in the evolutionary process adds an extra narrative dimension to the descriptions, which is particularly evident in improvisation 201[9] and helpful in understanding their evolution. Example 247[10] shows how imitation in the genetic model leads to much clearer structures, and thus it can be linked to Dawkins' premise of "copying-fidelity" [9] for a functioning evolution of memes.

---

[9] http://www.genimpro.net/index.htm?recordingID=201.
[10] http://www.genimpro.net/index.htm?recordingID=247.

4 Conclusion

Following the model of genetic improvisation, evolutionary processes are present on two levels, firstly individually in the invisible creative selection and secondly, as an evolution of the actually realized sound cells between the musicians (*interactive selection*). The latter can now be examined against the background of the theory of memetics as cultural evolution, based on the knowledge gained from the empirical application of the model.

It becomes clear that the model described here can be regarded as an instance of musical memetics, but this concept is expanded and specified in an improvisation-specific way. Contrary to the hierarchical structuring of a memetics of music [19, 32], which is derived from Schenker's concept, the lineages of the sound cells here form a co-adaptive network without any hierarchical structure. Jan emphasizes the necessary classification for a catalogue of musical memes – including the canon of European art music [19]. The genetic organization of the sound cells extends his proposal from one-dimensional classes to a multidimensional code that combines various generative parameters.

The perspective of memetics on the evolution of culture focuses not only on behaviours but also on products and thus derives its concept of *imitation* [2]. The *interaction* of the sound cells based on their specific development mechanisms is complementary to this on the genetic level, thus showing the deep structural code that first generates the artefacts – here the improvised sound. This also explains the fact that the genetic results of the empirical investigation do not coincide phenomenologically with the dominant principle of phonetic imitation, and thus appear partly counter-intuitive.

If we consider the algorithmic quality of an evolutionary process in the sense of Dennett [11], it should be possible to implement the principles of genetic improvisation in such a generative algorithm. However, it is questionable whether a fictitious genetic algorithm alone can produce aesthetically pleasing improvisations with the pursuit of the greatest possible fitness of the sound cells. In the creative selection of the individual players, which we can observe only as a result of the played sound cells, this principle can not be proven but suspected. On the contrary, in the second-order selection, which unfolds in the interaction of several improvisers, the interplay of relatively weak sound cells and the mechanisms of suppression of more powerful variants are found to be structurally particularly significant. Thus, artistic strategies and the Darwinian paradigm of the "survival of the fittest" seem to be opposing each other at first glance. However, this contrast only exists from a strict adaptionist point of view.

The generatively applied analysis methodology would result in a concrete implementation of an empirically grounded metacreation, but despite the algorithmic nature of evolution itself could not solve the problem of the fitness function. McCormack pointed out this problem of aesthetic selection, especially for artistic application [25]. The present study offers a "bottom-up approach" to find the code for an evaluation for this particular case [26]. The context of fitness and interaction introduced here may be a starting point for formalizing the problem, but it does not lead to generalization.

The function of the sound cell in the model of genetic improvisation marks the pivot point between the analytic and the generative and thus obtains the status of a "metagenerative system"[11], which unites both directions conceptually. Their genotype is a testimony to improvisational phylogenesis and the design of their sonic ontogenesis. For the retrospective, Dennett translates the process of analytic technique (*reverse engineering*) into a "hermeneutics of artefacts", but with the risk of implying an adaptationist meaning to every developmental step [10,11]. In contrast, the method of analysis of genetic improvisation works without such a teleological moment.

Within the metagenerative system of genetic improvisation, the distinction between the genotype and the phenotype of the sound cell marks the transition from the metagenarative of evolution to the generative of its ontogenesis, in which it brings to light the potentiality of its genetic character.

In order to contextualize the state of research achieved with this study, it is worthwhile revisiting a historical view of the field of genetics. Although the decoding of the human genome was an important milestone for genetics, it should not be confused with the comprehensive understanding of its mode of operation – despite intensive research efforts, there is still a long way to go. Similarly, the status of the sound cell genome can be interpreted. The mapping of the genome along with the revealed structures along empirical analysis is not the end of the process but must be considered as a starting point for further investigation. The elementary components of the synthetic theory of evolution may form a too rigid structure here to understand the dynamic character of the improvisation comprehensibly in all aspects. Thus, the model of genetic improvisation could follow the trends of evolutionary biology and be extended to include an epigenetics of improvisation, which models not only the sound cells but also their selection conditions.

The combination of sound cell genotypes and the resulting fitness forms a valuable body of implicit musical-improvisational knowledge, which can be applied to simulation and meta-creation using machine-learning methods. Just as a comprehensible correlation between the sound phenomenon and the sound cell dataset or even its phenomenological connection does not seem necessary for the analytical functioning, only the identical treatment of inputs and outputs is significant for the generative effectiveness. The structured access to the improvisational creative process opens up access to the pedagogical field in a new way in addition to the analytical and metagenerative. Machine-developed solution strategies often deviate from the intuitive ways of human decision-making and thus provide a starting point for reversing learning and discovering the creative in the machine.

---

[11] The term metagenerativity is closely related to the concept of musical metacreation [28], which, however, is restricted to a subsection of artificial creativity. The metagenerative level can be regarded as the superordinate regulator of individual generative processes [8].

References

1. Biles, J.: GenJam: a genetic algorithm for generating Jazz Solos. In: Proceedings of the 1994 International Computer Music Conference. University of Michigan Library (1994)
2. Blackmore, S.J.: The Meme Machine. Oxford University Press, Oxford (1999)
3. Boden, M.A.: The Creative Mind: Myths & Mechanisms. Basic Books (1991)
4. Bogdanov, D., et al.: ESSENTIA: an audio analysis library for music information retrieval. In: International Society for Music Information Retrieval Conference (ISMIR 2013), pp. 493–498 (2013)
5. Borgo, D.: Sync or Swarm: Improvising Music in a Complex Age. Continuum (2005)
6. Chomsky, N.: Aspects of the Theory of Syntax. MIT Press, Cambridge (1969)
7. Cock, P.J.A., et al.: Biopython: freely available Python tools for computational molecular biology and bioinformatics. Bioinformatics **25**(11), 1422–1423 (2009). https://doi.org/10.1093/bioinformatics/btp163
8. Dahlstedt, P.: Thoughts on creative evolution: a meta-generative approach to composition. Contemp. Music Rev. **28**(1), 43–55 (2009). https://doi.org/10.1080/07494460802664023
9. Dawkins, R.: The Selfish Gene. Oxford University Press, Oxford (1978)
10. Dennett, D.: The interpretation of texts, people and other artifacts. Philos. Phenomenol. Res. **50**, 177–194 (1990)
11. Dennett, D.: Darwin's Dangerous Idea: Evolution and the Meanings of Life. Simon & Schuster (1995)
12. Farris, J.S.: Estimating phylogenetic trees from distance matrices. Am. Nat. **106**(951), 645–668 (1972). https://doi.org/10.1086/282802
13. Ferretti, S.: On the modeling of musical solos as complex networks. Inf. Sci. (2016). https://doi.org/10.1016/j.ins.2016.10.007
14. Fitch, W.M., Margoliash, E.: Construction of phylogenetic trees. Science **155**(3760), 279–284 (1967). https://doi.org/10.1126/science.155.3760.279
15. Frieler, K., Lothwesen, K., Schütz, M.: The ideational flow: evaluating a new method for Jazz improvisation analysis. In: Proceedings of the 12th International Conference on Music Perception and Cognition and the 8th Triennial Conference of the European Society for the Cognitive Sciences of Music (2012)
16. Frieler, K., Pfleiderer, M., Zaddach, W.G., Abeßer, J.: Midlevel analysis of monophonic Jazz Solos: a new approach to the study of improvisation. Musicae Scientiae (2016). https://doi.org/10.1177/1029864916636440
17. Glaser, B.G., Strauss, A.L.: The Discovery of Grounded Theory: Strategies for Qualitative Research. Aldine Publishing Company (1967)
18. Hennig, W.: Phylogenetic Systematics. University of Illinois Press (1999)
19. Jan, S.: The Memetics of Music: A Neo-Darwinian View of Musical Structure and Culture. Ashgate (2007)
20. Johnson-Laird, P.N.: How Jazz musicians improvise. Music Percept.: Interdiscip. J. **19**(3), 415–442 (2002). https://doi.org/10.1525/mp.2002.19.3.415
21. Lartillot, O.: A musical pattern discovery system founded on a modeling of listening strategies. Comput. Music J. **28**(3), 53–67 (2004). https://doi.org/10.1162/0148926041790694

22. Lartillot, O., Ayari, M.: Comprehensive and complex modeling of structural understanding, studied of an experimental improvisation. In: Proceedings of the 12th International Conference on Music Perception and Cognition (ICMPC) and the 8th Triennial Conference of the European Society for the Cognitive Sciences of Music (ESCOM) (2012). http://cms.unige.ch/fapse/neuroemo/pdf/ArticleLartillotAyari.pdf

23. Lerdahl, F., Jackendorff, R.: A Generative Theory of Tonal Music. MIT Press, Cambridge (1996)

24. McAdams, S.: Perspectives on the contribution of timbre to musical structure. Comput. Music J. **23**(3), 85–102 (1999). http://www.mitpressjournals.org/doi/pdf/10.1162/014892699559797

25. McCormack, J.: Open problems in evolutionary music and art. In: Rothlauf, F. (ed.) Applications of Evolutionary Computing: EvoWorkshops2005, EvoBIO, EvoCOMNET, EvoHOT, EvoIASP, EvoMUSART, and EvoSTOC, Lausanne, Switzerland (2005)

26. Miłkowski, M.: Is evolution algorithmic? Minds Mach. **19**(4), 465–475 (2009). https://doi.org/10.1007/s11023-009-9170-6

27. Miranda, E. (ed.): A-Life for Music: Music and Computer Models of Living Systems. A-R Editions (2011)

28. Pasquier, P., Eigenfeldt, A., Bown, O., Dubnov, S.: An introduction to musical metacreation. Comput. Entertain. **14**(2), 2:1–2:14 (2017). https://doi.org/10.1145/2930672

29. Pressing, J.: Improvisation: methods and models. In: Sloboda, J.A. (ed.) Generative Processes in Music: The Psychology of Performance, Improvisation, and Composition, pp. 129–178. Oxford University Press, Oxford (1988)

30. Rahaim, M.: What else do we say when we say "Music Evolves?". World Music **48**(3), 29–41 (2006)

31. Saitou, N., Nei, M.: The neighbor-joining method: a new method for reconstructing phylogenetic trees. Mol. Biol. Evol. **4**(4), 406–425 (1987). http://mbe.oxfordjournals.org/content/4/4/406.short

32. Schenker, H.: Neue musikalische Theorien und Phantasien. Band 3: Der freie Satz. Universal-Edition (1935)

33. Tomlinson, G.: A Million Years of Music: The Emergence of Human Modernity. ZoneBooks (2015)

34. Trump, S.: The evolution of sound cells. pivot point for the analysis and creation of musical improvisation. In: Pasquier, P., Eigenfeldt, A., Bown, O. (eds.) Proceedings of the 4th International Workshop on Musical Metacreation (MUME 2016) (2016). http://musicalmetacreation.org/mume2016/proceedings/trump_the_evolution.pdf

35. Wallin, N., Merker, B., Brown, S. (eds.): The Origins of Music. MIT Press/A Bradford Book (2000)

Controlling Self-organization in Generative Creative Systems

Jonathan Young[1]([✉])(iD) and Simon Colton[2,3]([✉])(iD)

[1] EECS, Queen Mary University of London, London, UK
j.a.young@qmul.ac.uk
[2] SensiLab, Faculty of IT, Monash University, Melbourne, Australia
[3] Game AI Group, EECS, Queen Mary University of London, London, UK
s.colton@qmul.ac.uk

Abstract. We present a new tool which simulates the development of Artificial Chemistries (AChems) to produce real-time imagery for artistic/entertainment purposes. There have been many such usages of complex systems (CSs) for artistic purposes, but deciding which parameters to use for such unpredictable systems can lead to a feeling of lack of control. For our purposes, we struggled to gain enough control over the AChem real-time image generation tool to accompany music in a video-jockeying application. To overcome this difficulty, we developed a general-purpose clustering approach that attempts to produce sets of parameter configurations which lead to maximally distinct visualisations, thus ensuring users feel that they have influence over the AChem when controlled with a suitable GUI. We present this approach and its application to controlling the development of AChems, along with the results from experiments with different clustering approaches, aided by both machine vision analysis and human curation. We conclude by advocating an overfitting approach supplemented by a final check by a designer, and discuss potential applications of this in artistic and entertainment settings.

Keywords: Artificial Chemistry · Artificial life · Generative creativity · Unsupervised learning · Guided Self-organization

1 Introduction and Motivation

Guided Self-organization (GSO) is a field of research concerned with establishing a standard information-theoretic framework for guiding self-organization in complex systems (CSs) [21]. Self-organization is generally accepted as a process which results in rich and structured dynamics from a set of simple evolution rules. In biological and man-made systems, the objective of these dynamics is functionality [4]. Guidance in GSO has been characterized as a mid point between completely designed solutions and completely uninfluenced ones [21]. Hence the field of GSO is predicated on the fact that CSs are unpredictable enough that they require guidance to attain certain goals, especially if those systems have

J. Romero et al. (Eds.): EvoMUSART 2020, LNCS 12103, pp. 194–209, 2020.
https://doi.org/10.1007/978-3-030-43859-3_14

complex multi-modal parameter spaces. For example, there may be many possible sets of rules that govern behaviour, many possible combinations of values for numerical parameter settings, or a combination of the two.

CSs, such as artificial life (AL) simulations, have many qualities that are desirable in generative creative applications [12]. Such qualities include the generation of complexity [10], the emergence of unexpected phenomena from simple interactions [20], and the convergence-resistant generation of novel phenomena [9]. For example, the usage of distributed systems such as artificial ecosystems [8] and cellular automata [2] have been popular in generative art and music because of their ability to autonomously generate complex and novel structures or dynamics from simple rules. Typically the state of the system at each time step is mapped to some visual or audible output. Designers will often attempt to guide the system's self-organization towards emergent dynamics and structures that are aesthetically pleasing. For example, interactive evolution has been used to evolve the rules of cellular automata and virtual ecosystems [16]. Designers may pre-evolve a set of individuals to be sampled from, thus creating the initial population of the interactive evolutionary algorithm [8,16]. Additionally, some works allow users to modulate the parameters of the system in order to guide self-organization in real time. For example, the Eden system uses sensors to track user movement in order to modulate parameters [19]. The Meniscus system allows users to adjust the water level in the environment [7]. The user or designer often judges the value of generated output because determining how aesthetically pleasing or creative it is can be difficult to quantify computationally or information-theoretically such as in GSO. However, the use of human judges is problematic because it is time consuming to manually compare many generated solutions. Opinions on aesthetics, creativity or interestingness are also highly subjective. So it is difficult for a researcher to understand what parameters give universally optimal solutions when a consensus is lacking in the judgement of said solutions [15].

Common methods for guiding self-organization in generative systems can be problematic in systems with unpredictable parameters spaces. Many phenomena of interest may never emerge if such a parameter space is restricted, but redundant phenomena may frequently emerge if the space is explored without restriction. In general the unpredictability of parameter choices when applied to CSs can lead to a feeling of lack of control for the user.

We developed a new video-jockeying (VJ) tool which features a visual representation of an Artificial Chemistry (AChem) that is updating its state in real time. Control over the AChem is necessary if the user wants to influence it to effectively accompany music. We tried to simply allow certain parameters to be freely manipulated by the user. Additional modes of interaction were introduced, such as allowing the addition of atoms using the mouse. However, none of these methods afforded us a sufficient sense of control over the inherently unpredictable nature of the AChem. To the best of our knowledge, no standardized procedure has been proposed that gives users a sense of control over the unpredictable parameter spaces of CSs. Such a procedure could help to allow many more CSs or AL simulations to be controlled for the purposes of generative art.

We propose a procedure that attempts to maximize the heterogeneity of the set of emergent phenomena a CS produces by limiting parameter space. Users may therefore be afforded a feeling of control over self-organization by observing a sufficient change in emergent phenomena when they switch between configurations from this limited parameter space. Hence, maximizing the perception of heterogeneity minimizes the perception of redundancy. The procedure first involves randomly generating many parameter configurations for the AChem. Secondly, clustering and computer vision are used to narrow down configurations by analysing the resulting AChems. Thirdly, images of AChems with configurations representing the cluster centroids have their pair-wise similarity rated by human judgement and indistinguishable clusters are removed. Finally sets of clusters are reduced to desired subset sizes by taking the lowest similarity subset. The set of parameter configurations with the lowest similarity rating is made available for users to choose from in the VJ application. At any given time the user can choose a configuration and observe the AChem change in real time. Suitable clustering methods are compared and we advise that an overfitted model tempered by human judgement may tend to produce lower average similarity than an underfitted model. A rigorous evaluation of clustering methods is beyond the scope of this study.

The procedure is designed to be general-purpose. So the end user, CS or AL simulation, clustering approach, artistic output medium and the computational analysis of said output are theoretically interchangeable. For example, A designer might integrate the resulting configurations into an autonomous generative system, or have the audience switch between them in an interactive system. The procedure could also cater for the generation of distinct sets of static imagery. The procedure also avoids restricting the parameter space based on presupposed subjective notions of interestingness or creativity.

The paper proceeds as follows: AChems are introduced and discussed in Sect. 2, Sect. 3 outlines the general-purpose procedure, Sect. 4 details the AChem that the procedure is applied to, Sect. 5 details the experimental application of the procedure to the AChem, Sect. 6 outlines the results of the experiment and Sect. 7 concludes the study and discusses future work.

2   Artificial Chemistries

Artificial Chemistries (AChems) are computational systems which emulate aspects of real chemistry. AChems are often used to attempt to answer AL research questions, such as the processes from which biological life may have emerged [6], and routes to open-ended dynamics [11]. AChem research abstracts from specific examples of real natural processes in an attempt to investigate generic principles of life and its origin which are not historically contingent or bound to a specific substrate [6]. This approach to research has its roots in

AL [17]. Many AChems have characteristics which limit their ability to produce open-ended dynamics [24]. Hence, innovations in the field are often focused on removing these limits.

AChems can be defined by a triple (S, R, A) where S is the set of possible molecules, R is the set of rules for molecular interactions, and A is the algorithm that describes the environment in which R can be applied to S. The approach to representing molecules in S can be characterized as symbolic, structured or sub-symbolic. Symbolic molecular representations have no internal structure and are usually represented by a symbol. All molecules must therefore be explicitly defined [26]. In contrast, structured molecular representations have some internal structure, such as a set of atoms bound together in a specific arrangement. In this approach, molecules (and possibly reactions) can emerge from combinations of lower-level components [14]. Sub-symbolic molecular representations have an internal structure defined by sub-atomic components analogous to those found in nature [11]. However, both the structured and subsymbolic approaches can result in AChems that are difficult to analyse and computationally expensive to simulate.

Interactions in R between molecules can be explicitly defined, implicitly emerge from properties of molecules, or a combination of the two. Explicit interaction schemes require any interactions to be manually defined. Some AChems use a grammar-like notation to define an explicit set of reactions [26]. Implicit interaction schemes can be algorithmically applied to any possible combination of molecules. So interactions emerge from the properties of the molecules or atoms themselves, including those that are not part of the original specification [1]. Some AChems have a combination of explicit and implicit reaction schemes [14]. This approach allows potential for emergent novel interactions and therefore open-ended dynamics. However, the resulting emergent dynamics can be unpredictable and highly sensitive to minor perturbations.

The algorithm A that represents the environment can be characterized as well-mixed, continuous space or discrete space. In a well-mixed environment all molecules can interact at any given time, so reactants have no location [1]. In continuous or discrete spaces, molecules typically only interact if spatially proximate. Spatial partitioning may be an important stage in the progression towards the origin of life, so the subject is a focus in various AChem studies [26]. Discrete space is computationally cheap to represent although less able to represent spatial nuances when compared to continuous spaces.

3   A Procedure for Increasing Heterogeneity

The procedure for attempting to limit parameter space such that the heterogeneity of the set of emergent phenomena is maximized first involves extracting an appropriate subset of parameter configurations from a set of available ones. Let C^{min} be the former and C the latter, where $C^{min} \subset C$, $c^{min} \in C^{min}$, $c \in C$ and $a \leq |C^{min}| \leq b$ for some user-defined range a and b. Each c^{min}, once applied to the CS, should elicit phenomena that are maximally qualitatively different to each other. Hence, to find c^{min} we need to minimize some measure of similarity. Let $\alpha(X, f)$ be a measure of average similarity between all pairs of configurations within some set of configurations X, where symmetric function $f(x, y)$ assesses the similarity (computationally or by human judges) between the pair of configurations $\{x, y\} \in X$ once each of them is applied to the CS. $\alpha(X, f)$ is calculated such that

$$\alpha(X, f) = \sum_{\{x_1, x_2\} \in [X]^2} \frac{f(x_1, x_2)}{|[X]^2|} \tag{1}$$

where $[X]^y$ denotes all subsets in set X of size y (note that each subset is distinct as are the elements therein). $f(x, y)$ could typically give a real number between 0 and 1. The procedure can be characterized as an attempt to approximately construct C^{min} such that

$$\forall C^{ab} \in \bigcup_{i=a}^{b} [C]^i, \alpha(C^{min}, f) \leq \alpha(C^{ab}, f) \tag{2}$$

Constructing C^{min} and applying c^{min} to the CS allows its state space to be explored, whilst omitting redundant c that produce overly similar phenomena upon application. This enables a subset of parameter space that produces heterogeneous phenomena to be explored.

3.1   Clustering Methods for Redundancy Checking

In some cases it is impractical for a human judge to go through every c and assess its redundancy. We consider redundant configurations to be those that produce indistinct emergent phenomena in the CS. Clustering methods should be used to eliminate any redundant c from C because they can identify distinct groups in the data, thus allowing configurations that produce distinct phenomena to be picked and others to be disregarded. The clustering algorithm of choice must operate on data that reflects some perceived characteristic of the CS. If the CS is given a visual or audible mapping, then a data point could be some differentiable quantity x^m calculated by a computer vision or machine listening analysis respectively of the CS with configuration x. Let $g_i(X)$ be the set of representative configurations attained by applying clustering method i to a set of configurations X, where $x \in g_i(X)$, each x^m is closest to each centroid and $g_i(X) \in \{g_1(C), ..., g_n(C)\}$. Hereafter, $g_i(X)$ will be referred to as centroid configurations. Constructing $g_i(X)$ ensures that redundancy is reduced because $\alpha(g_i(X), f^c) \leq \alpha(X, f^c)$ where $f^c(x, y)$ is the distance between x^m and y^m, and $\{x, y\} \subset g_i(X)$. The aim of selecting centroid configurations is to ensure that each configuration closely represents the key identifying features reflected by each cluster. Selecting a configuration from each cluster for which x^m is furthest from the others is a valid alternative approach that could further reduce redundancy. However, such an approach may disregard patterns found in the data by the clustering algorithm and allow configurations to be selected that do not contribute to these patterns. Configurations such as these could be those that introduce stochasticity into the CS and its subsequent analysis. Examples of testing multiple clustering methods could include applying multiple clustering algorithms (such as k-means or self-organizing maps), multiple values of k, or multiple methods for determining k computationally, where k is the number of clusters or groups for the algorithm in question.

3.2 Human Redundancy Checking

Human judgement can be used instead of clustering methods to attain C^{min} if the size of C is small enough, but human judgement should still be used to assess clustering methods for eliminating redundancy because they are unlikely to perfectly match human perception. The designer should iterate through all pairs of configurations $\{x_1, x_2\} \in [X]^2$, where X is some set of configurations. If x_1 and x_2 produce indistinguishable phenomena, randomly choose one and remove it from X. Let $h(X)$ be the resulting post-elimination set of configurations. $\alpha(h(X), f) \leq \alpha(X, f)$ implicitly because pairs that are too similar are being eliminated.

3.3 Bounds Filtering

After human judgement with or without clustering methods has been used to eliminate redundant configurations, the remaining set of configurations needs to fit within the user-defined range a and b for it to be useful. If the remaining set of configurations is too large, then the subset of size b with the lowest average similarity should be chosen. Let $\eta(X)$ be the set of configurations attained by filtering the remaining configurations X such that:

$$\eta(X) = \begin{cases} \varnothing, & \text{if } |X| < a \\ X, & \text{if } a \leq |X| \leq b \\ min(X), & \text{if } |X| > b \end{cases} \tag{3}$$

$$\text{where: } \forall X^b \in [X]^b, \alpha(min(X), f^r) \leq \alpha(X^b, f^r), \tag{4}$$

$min(X) \in [X]^b$, and $f^r(x, y)$ is an explicit rating of similarity provided by the designer for configurations x and y once applied to the CS and perceived by the designer. Calculating $\alpha(X, f^r)$ requires a designer to iterate through all pairs $x_1, x_2 \in [X]^2$ in random order and provide a rating $f^r(x_1, x_2)$ for some set of configurations X. If rating pairs from multiple clustering methods, all pairs should be intermixed and randomly ordered to ensure fatigue does not disproportionately affect ratings for particular clustering methods.

3.4 Combining Steps

C^{min} can be constructed with all the aforementioned methods. If only human judgement is being used to construct C^{min}, then $C^{min} = \eta(h(C))$. If human judgement and a single clustering method is being used to construct C^{min}, then

$C^{min} = \gamma_1(C)$ where $\gamma_i(X) = \eta(h(g_i(X)))$ for some set of configurations X. However, if the designer is assessing multiple clustering methods, let $C^{min} = \gamma_{min\,\alpha}(C)$ such that $\gamma_{min\,\alpha}(C)$ is the lowest similarity set out of all clustering methods to ensure higher perceptual heterogeneity:

$$\forall i, \alpha(\gamma_{min\,\alpha}(C), f^r) \leq \alpha(\gamma_i(C), f^r) \tag{5}$$

Figure 1 shows a summarized version of the procedure.

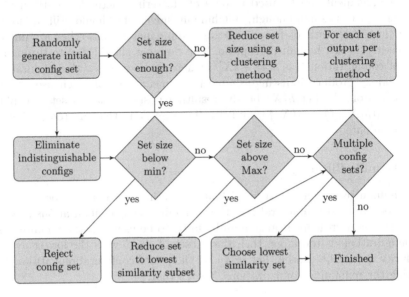

Fig. 1. A summary of the procedure as a flow chart.

3.5   Fitting

Underfitting and overfitting can occur in clustering methods if k is too low or too high respectively (where k is the number of clusters). For example, underfitting increases the risk of grouping configurations that produce distinct phenomena. Conversely, overfitting increases the risk of separating configurations that produce similar phenomena which are only trivially different due to chance. For example, two configurations may produce a very similar characteristic overall pattern in an AChem but slightly different positioning of atoms due to random initial positioning. k is proportional to the number of divisions in the data and therefore correlates with the variance in distances between pairs of centroids. Hence, certain subsets of clusters from an overfitted model may be far more distant in terms of their centroids than those from an underfitted model. Hence, overfitting can potentially be tempered by removing similar centroid configurations via redundancy checking and bounds filtering (Sects. 3.2 and 3.3 respectively), as long as centroid distances correlate with similarity ratings. So whilst it is advisable to test multiple k or methods for determining k when applying the procedure to avoid overfitting and underfitting, it is preferable to overfit and temper using human judgement than to underfit. However, simply using a very high k will provide an impractical rating task for the designer.

4   Application for Controlling Artificial Chemistries

To test the procedure, an AChem simulation was implemented in Unity3D. It is a structured AChem with four possible chemical elements per atom and an explicit interaction scheme in a two dimensional lattice of size 80×80. The lattice has a periodic boundary, which means that space repeats itself. Hence, atoms and bonds can pass through any boundary of the lattice and re-appear on the opposite side. Periodic boundaries are used in some AChem simulations [18]. The AChem takes a set of atomic bond forming or breaking reaction rules which constitutes a configuration c for the AChem. Each atom can have a maximum of 10 bonds and each reaction rule is a tuple

$$\rho = (e_1, e_2, m, b_1, b_2, c_1, c_2, p) \tag{6}$$

where e_i is the required element of atom i, m is true or false if the reaction makes or breaks a bond respectively between both atoms and b_i is one of four required states of atom i: unbonded, bonded to an atom of element c_i, not bonded to an atom of element c_i or any bonding state. Note that there can only be one bond between any two atoms. p is the probability that the reaction will be attempted, at which point the update algorithm searches for atoms that the reaction rule can be applied to. The update algorithm is explained in more detail below.

The update algorithm for the simulation attempts to apply each of the rules to any atoms within a 6×6 sized section of the lattice starting at a randomly chosen position. This process repeats $round(80^2/6^2)$ times per update cycle, thus ensuring approximately the entire lattice is updated at every cycle [23]. The reaction site size and lattice size were chosen after some testing because finding optimal values is beyond the scope of this study. Note that this approach allows bonds to form across the periodic boundary. This 6×6 section of the lattice will be referred to as the reaction site hereafter. When applying the rules to a section of the lattice, each reaction rule is iterated over. Each rule is attempted with probability p (see Eq. 6). When a reaction is attempted, it will try to randomly select two atoms from the reaction site that meet the conditions of the reaction rule (shown in Eq. 6). This involves iterating through all atoms in the reaction site, finding pairs of atoms that the rule can be applied to, randomly selecting a pair, and attempting the bond creation or decomposition provided there is an available pair. A pair is invalid if the atoms are not of the elements e_1 and e_2, or if they do not have the requisite bond statuses set out by b_1, b_2, c_1, c_2, or if the rule breaks or makes a bond according to m but the atoms are bonded or not bonded already respectively.

In addition to attempting to apply the reaction rules, all bonds that connect to any atoms in the reaction site are broken if the bonded atoms are over

6 cells apart in any dimension. Each atom in the reaction site will also randomly move to a cell in its moore neighbourhood, which is the cell the atom is occupying and all of its adjacent cells including diagonals. Each atom in the reaction site also stochastically moves towards any atoms it is bonded with. Let $p_i = |d_i| / \sum_{j=1}^{2} |d_j|$ be the probability of this movement in each dimension i where $d_i = p_i^2 - p_i^1$, p_i^1 is the position of the atom making the stochastic move and p_i^2 is the position of the bonded atom. Let m_i be the movement of the bonded atom that is generated such that

$$m_i = \begin{cases} sign(d_i), & \text{if } \mathcal{R} < p_i \\ 0, & \text{if } \mathcal{R} \geq p_i \end{cases} \tag{7}$$

where $\mathcal{R}$ is a randomly generated real number between 0 and 1 and $sign(x)$ returns 1, 0 or -1 if number x is positive, zero or negative respectively. The probabilities are higher in dimensions where the bonded atoms are more distant. Hence, the atom is more likely to move towards its bonded atom in those dimensions. The explicit interaction scheme ensures computational cheapness, but structured molecular representations allows some structure and behaviour to emerge. The procedure we proposed is designed to allow us to control an inherently difficult to predict system, and a symbolic AChem with explicit rules might be overly predictable. An explicit rule-based interaction scheme was implemented to demonstrate how even a highly dimensional and interdependent configuration space can be controlled by the procedure proposed in this study.

5   Application to Realtime Animation Generation

We randomly generated 10000 parameter configurations each consisting of 20 randomly generated reaction rules for the AChem described in Sect. 4 above. This makes up the set of configurations C. Each AChem was initialized by filling a third of the lattice ($round(80^2/3)$) with atoms, an amount that was chosen after some initial testing. Each atom was initialized with a randomly assigned element and lattice position. Each AChem was updated for 1000 update cycles, after which an image of the AChem's final state is generated and stored. Each pixel of the image corresponds to a position in the lattice. Each atom is assigned a hue that is unique to its element. $hue(i) = e^l(i)/n_{el}$ where $e^l(i)$ is the element of atom i, n_{el} is the total number of elements and $e^l(i) \in \{0, ..., n_{el} - 1\}$. Empty lattice positions are rendered as black. Examples of these images can be seen in Fig. 4. Then each image was fed into ResNet50 [13], after which the output layer vector was saved. Let $\mathbf{o}$ be the set of these vectors for each AChem where $\mathbf{o} \in O$. A probabilistic representation that reflects visual aspects of the AChem's final state is also stored. Let $\mathbb{P}_{ij} = \epsilon_{ij}/\epsilon$ be the joint distribution of bonded elements where ϵ_{ij} is the number of bonds that connect an atom of element i to another of element j, and ϵ is the total number of bonds. Each distribution $\mathbb{P}_{ij}$ is reshaped into a vector and stored in set R. Both $\mathbf{o}$ and $\mathbb{P}_{ij}$ can be considered aspects of the AChem that can be visually perceived. For example, bonded atoms which have a colour corresponding to their element will be closer together in space. So $\mathbb{P}_{ij}$ is a representation of the resulting emergent spatial and visual structure of the AChem. $\mathbf{o}$ is the output of a neural network which was trained to be able to classify images in a similar way to humans. So the output vector should contain information about the images of AChems which can be humanly perceived. The ResNet50 vectors are calculated after the image has been generated.

K-means clustering was then applied to R and O. We tried different ways of determining a suitable k for both sets. The Calinski-Harabasz indices [3], Davies-Bouldin indices [5], gap criterion values [25] and silhouette values [22] were used for evaluating clusters generated by the k-means algorithm for all k values 1 to 40. A specific explanation of each method for evaluating k is beyond the scope of this study. Each clustering method is represented by $g_i(C)$ of which there are 8 in total (4 ways of determining k applied to R and O) where $g_i(C)$ are the centroid configurations and each measurement x^m for each configuration $x \in g_i(C)$ is either a $\mathbb{P}_{ij}$ or $\mathbf{o}$. These parts of the experiment also make up Sect. 3.1 of the procedure described above.

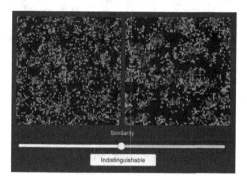

Fig. 2. The rating tool. The designer clicks and drags the slider to rate the similarity of the pair of configurations, or they can click the button below if they are deemed indistinguishable.

We implemented a tool that allowed us to evaluate clusters. The tool shows images for each pair of configurations within each $g_i(C)$ where $a \leq |g_i(C)|$, and allows us to rate their level of similarity as a real number between 0 and 1 using a slider. This rating is represented by f. If a pair of configurations are indistinguishable, we can press a button which causes one configuration out of the pair to be randomly chosen and removed. A screenshot of the tool can be seen in Fig. 2. Once this process was complete, $min(g_i(C))$ was calculated for each remaining $g_i(C)$ where $|g_i(C)| > b$. This entire process is represented by $\gamma_i(C)$ because indistinguishable configurations are being removed, pairs of configurations are being evaluated and the range represented by a and b is being ensured. a and b in this experiment are 8 and 12 respectively. Note that pairs of configurations may repeat if different clustering methods result in some of

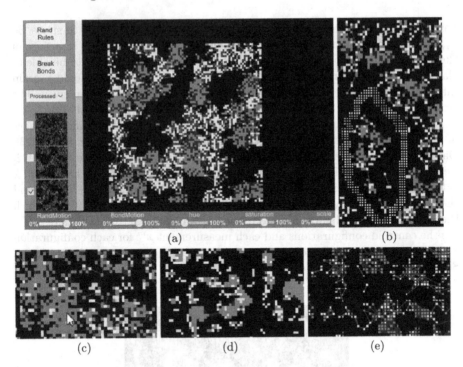

Fig. 3. (a) A screenshot of the VJ tool with the AChem visualized in the centre. The leftmost panel contains configuration selection, bond break and configuration randomization buttons. The bottommost panel contains sliders that control visual style and atom movement. (b) A user-drawn barrier (black and white outline) in the AChem. (c) Red atoms being added to the AChem near the mouse pointer. (d) The AChem with zero probability of random atom motion, which causes atoms to clump together. (e) The AChem with atom scale set to 60%. (Color figure online)

the same configurations. Pairs that have already been rated are skipped over by the tool and the original rating is reused. All pairs in all sets generated by each clustering method are intermixed and shown to the designer in a random order. The order of images within the pair when shown to the designer is also random. To avoid fatigue, ratings were done for 30 min at a time with a minimum of 10 min to rest in between. These parts of the experiment make up Sects. 3.2 and 3.3 of the procedure given above. Finally our desired set C^{min} was chosen as the clustering method that produces the lowest similarity set $\gamma_{min\,\alpha}(C)$ after the entire procedure is completed, the construction of which is fully explained in Sect. 3.4 of the above procedure. This set of configurations is then integrated into a VJ tool in which the user can select each configuration to change the set of reactions in an AChem updating in real time. When the user selects a configuration, all chemical bonds are broken to aid in causing the AChem to converge to the desired state. Figure 3(a) shows a screenshot of the tool in which the set of configurations can be chosen from on the leftmost panel. This panel shows the set of images originally analysed by ResNet50 that correspond to the set of configurations. Each image can be clicked on to apply its corresponding configuration to the AChem shown in the centre of the tool as it updates in real-time. The visual representation of the AChem that the configurations are applied to is identical to the images generated for analysis by ResNet50, except the atoms are enlarged and the bonds between them are visualized as lines. There are many other features of this tool which were initially implemented to provide users with a sense of control over the AChem prior to integrating C^{min}. For example, the user can also alter aspects of atom movement and the visual style of the AChem using continuous sliders on the bottommost panel. They can cyclically shift the hue and alter the saturation of the atoms with sliders akin to those used in image manipulation software. There is also a slider that changes the atom scale, the values of which correspond to the width of an atom i.e. the width of a lattice cell at its largest. Atom scale is maximal (100%) by default, but an example of the AChem with 60% atom scale can be seen in Fig. 3(e). The probability of random atom motion and stochastic atom motion towards bonded atoms (described in Sect. 4) can also be manipulated via sliders. These probabilities are 1 by default, but an example of the AChem with no probability of random atom motion can be seen in Fig. 3(d). All the bonds in the AChem can be broken at will, or a random configuration can be chosen using additional buttons on the leftmost panel. Atoms can be added or removed under the pointer by holding the left or right mouse button respectively and dragging. The number keys 1–4 can be used to select the element that added atoms should be. An example of atoms being added to the AChem can be seen in Fig. 3(c). A barrier can be drawn by holding shift and clicking and dragging the mouse across the screen. This barrier annihilates atoms that come into contact with it. An example of such a barrier can be seen in Fig. 3(b).

6 Results

Table 1 shows the configuration set sizes and average similarities at every stage of the application of the procedure described in Sect. 5 above. The initial number of centroid configurations per method is the second column, the number of these configurations left after redundancy removal (Sect. 3.2) is the third column, the average similarity after indistinguishable configurations are removed is the fourth column and the final average similarity and number of configurations after the set undergoes bounds filtering (Sect. 3.3) is the fifth and sixth column respectively. The gap criterion for determining k for k-means clustering resulted in the lowest final average similarity when applied to R. This set of configurations was integrated into the VJ tool and can be considered C^{min}. Figure 4 shows the original stored images for these configurations, and corresponding images of the AChem in the VJ tool 15 s after each configuration is applied in sequence. The gap and silhouette methods for determining k result in low similarity ratings generally. They also result in high numbers of initial configurations or clusters, many of which are removed in the process of redundancy checking or bounds filtering. The gap criterion is the only method of determining k which does not fail to converge when applied to O. Hence, the last three columns in Table 1 are blank for the remaining methods that operate on O, because after redundancy checking there are no clusters left with which to continue the procedure. This failure to converge is likely to be related to the high-dimensionality of the output vectors for ResNet50 which have a length of 1000. However, a full discussion of this problem is beyond the scope of this study. The Calinski-Harabasz and Davies-Bouldin methods for determining k when applied to R result in small sets of configurations to begin with and high average similarities. After redundancy checking, the number of configurations is too small to meet the minimum user-defined bound a in these cases. Hence, the two last columns in Table 1 are blank for these methods because the procedure cannot be continued.

Table 1. The results for all stages of the procedure. The left most column is the clustering method. The name of the method begins with either "Prob" or "ResNet" to signify the use of probability distributions or ResNet50 output. The name of the method ends with method for determining k (e.g. "gap" or "CalinskiHarabasz").

Method	$|g_i(C)|$	$|h(g_i(C))|$	$\alpha(h(g_i(C)), f^r)$	$\alpha(\gamma_i(C), f^r)$	$|\gamma_i(C)|$
ProbCalinskiHarabasz	10	4	0.54		
ProbDaviesBouldin	10	4	0.63		
Probsilhouette	36	22	0.38	0.29	12
Probgap	40	21	0.37	0.13	12
ResNetCalinskiHarabasz	3	0			
ResNetDaviesBouldin	2	0			
ResNetsilhouette	2	0			
ResNetgap	39	18	0.38	0.29	12

6.1   Fitting

The silhouette method and gap criterion approaches result in lower final average similarities and higher initial numbers of configurations or clusters because they may overfit. This may be why there is such a reduction in average similarity after the lowest average similarity subset of configurations of size b is selected (see Sect. 3.5). Overfitting is also reflected in the amount of indistinguishable configurations that needed to be removed during redundancy checking for these methods due to being separated into different clusters despite producing similar measurements (see Sect. 3.5). This interpretation is bolstered by the fact that we found many pairs of images of configurations that looked almost identical. However, there are aspects of the data itself that may cause many indistinguishable or highly similar centroid configurations to produce measurements which are highly distinct. For example, bonded element distributions $\mathbb{P}_{ij}$ may be distant

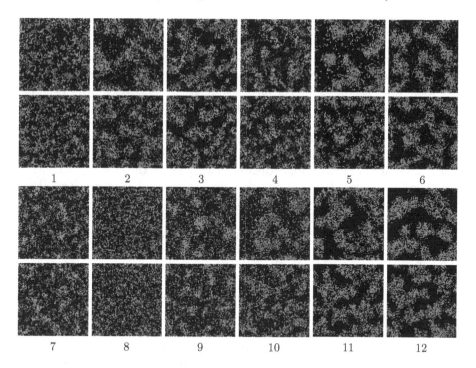

Fig. 4. Images of the finally chosen configuration set, and the AChem in the VJ tool 15 s after each configuration is applied in sequence. The top image within each pair is the original image generated in the experiment, and the bottom one is of the VJ tool after the corresponding configuration is applied.

in euclidean space when treated as vectors (see Sect. 5), but their corresponding images may be judged as indistinguishable or highly similar. For example, two AChems may look extremely similar because they would have a near identical distribution of bonded elements if some of the elements were swapped. However, the vectors of their corresponding bonded element distributions $\mathbb{P}_{ij}$ will be far apart because they reflect the probability of bonds between specific elements. We also found many pairs of images of configurations that looked almost identical in structure except for the distribution of colours. Colours correspond to elements, so this finding bolsters the aforementioned interpretation of the data.

The Calinski-Harabasz and Davies-Bouldin methods result in higher average similarity ratings. This could be a result of underfitting. The smaller amount of clusters group together data that might be deemed distinct otherwise (see Sect. 3.5). Hence, the resulting centroid configurations produce measurements that are not only similar to each other, but also an approximate average over more distinct phenomena. The similarity ratings probably reflect this interpretation. A specific interpretation of why each clustering method gave these results is beyond the scope of this study.

7 Conclusion and Future Work

We proposed and demonstrated a general-purpose procedure that affords users an increased sense of control over a real-time visual representation of an AChem in a VJ tool. However, the tool needs further evaluation and development with professional VJ practitioners to determine the requirements for its real-world applicability. The results of applying the procedure supported an intuition that clustering methods that overfit to the data generated by configurations are preferable because they can then be filtered through human judgement to attain a set of configurations that produces an approximately maximally distinct set of phenomena, thus affording a sense of control by switching between said configurations. However, during the rating task it was difficult to give a rating of similarity that took into account all features of the image. Hence, it became clear that individual raters may subjectively prioritize different features of images. A participant study should be conducted to ensure that ratings provided by the designer and the general public significantly agree. Otherwise the ratings may not be reliable enough to induce a feeling of control in a general userbase. This lack of inter-rater reliability could hinder the future evaluation and development of the VJ tool. Additionally, computational similarity assessments may exist which can replace human ratings if they are correlated. Configurations that produce more complex phenomena may be ignored during clustering unless complexity is computationally determined and used as a weighting in a weighted clustering algorithm. Alternatively, a human rating of interest could weight configurations so that ones that produce interesting phenomena are less likely to be removed during bounds filtering (Sect. 3.3). This may aid in engaging the intended users and audience of the VJ tool. The version of the proposed procedure we applied uses centroid configurations during redundancy checking (Sect. 3.1). Despite our justifications for this approach, selecting a configuration per cluster with measurements furthest from each other may be a superior approach that could also be tested. The clustering methods themselves could also be more rigorously evaluated and compared.

Acknowledgements. We would like to thank three anonymous reviewers for helpful comments. Jonathan is funded by the EPSRC and AHRC Centre for Doctoral Training in Media and Arts Technology.

References

1. Banzhaf, W.: Self-organization in a system of binary strings. In: Artificial Life IV (1994)
2. Beyls, P.: The musical universe of cellular automata. In: Proceedings of International Computer Music Conference (1989)
3. Caliński, T., Harabasz, J.: A dendrite method for cluster analysis. Commun. Stat.-Theory Methods **3**(1), 1–27 (1974)
4. Claudius, G.: Complex and Adaptive Dynamical Systems: A Primer. Springer, Heidelberg (2008). https://doi.org/10.1007/978-3-540-71874-1
5. Davies, D.L., Bouldin, D.W.: A cluster separation measure. IEEE Trans. Pattern Anal. Mach. Intell. **2**, 224–227 (1979)

6. Dittrich, P., Ziegler, J.C., Banzhaf, W.: Artificial chemistries-a review. Artif. Life **7**(3), 225–275 (2001)

7. Dorin, A.: Meniscus. In: Taylor (ed.) Experimenta: House of Tomorrow Catalogue. Experimenta Media Arts, Australia (2003)

8. Dorin, A.: The virtual ecosystem as generative electronic art. In: Raidl, G., et al. (eds.) EvoWorkshops 2004. LNCS, vol. 3005, pp. 467–476. Springer, Heidelberg (2004). https://doi.org/10.1007/978-3-540-24653-4_48

9. Dorin, A.: Artificial life, death and epidemics in evolutionary, generative electronic art. In: Rothlauf, F., et al. (eds.) EvoWorkshops 2005. LNCS, vol. 3449, pp. 448–457. Springer, Heidelberg (2005). https://doi.org/10.1007/978-3-540-32003-6_45

10. Dorin, A.: Chance and complexity: stochastic and generative processes in art and creativity. In: Proceedings of the Virtual Reality International Conference: Laval Virtual (2013)

11. Faulconbridge, A., Stepney, S., Miller, J.F., Caves, L.S.: RBN-world-the hunt for a rich AChem. In: ALife (2010)

12. Galanter, P.: What is generative art? Complexity theory as a context for art theory. In: GA2003-6th Generative Art Conference (2003)

13. He, K., Zhang, X., Ren, S., Sun, J.: Deep residual learning for image recognition. In: Proceedings of the IEEE Conference on Computer Vision and Pattern Recognition (2016)

14. Hutton, T.J.: Evolvable self-replicating molecules in an artificial chemistry. Artif. Life **8**(4), 341–356 (2002)

15. Jordanous, A.: A standardised procedure for evaluating creative systems: computational creativity evaluation based on what it is to be creative. Cogn. Comput. **4**(3), 246–279 (2012). https://doi.org/10.1007/s12559-012-9156-1

16. Kowaliw, T., McCormack, J., Dorin, A.: An interactive electronic art system based on artificial ecosystemics. In: 2011 IEEE Symposium on Artificial Life (ALIFE) (2011)

17. Langton, C.G., et al.: Artificial life (1989)

18. Madina, D., Ono, N., Ikegami, T.: Cellular evolution in a 3D lattice artificial chemistry. In: Banzhaf, W., Ziegler, J., Christaller, T., Dittrich, P., Kim, J.T. (eds.) ECAL 2003. LNCS (LNAI), vol. 2801, pp. 59–68. Springer, Heidelberg (2003). https://doi.org/10.1007/978-3-540-39432-7_7

19. McCormack, J.: Eden: an evolutionary sonic ecosystem. In: Kelemen, J., Sosík, P. (eds.) ECAL 2001. LNCS (LNAI), vol. 2159, pp. 133–142. Springer, Heidelberg (2001). https://doi.org/10.1007/3-540-44811-X_13

20. McCormack, J., Dorin, A.: Art, emergence and the computational sublime. In: Proceedings of Second Iteration: A Conference on Generative Systems in the Electronic Arts. CEMA, Melbourne (2001)

21. Prokopenko, M.: Guided self-organization (2009)

22. Rousseeuw, P.J.: Silhouettes: a graphical aid to the interpretation and validation of cluster analysis. J. Comput. Appl. Math. **20**, 53–65 (1987)

23. Suzudo, T.: Spatial pattern formation in asynchronous cellular automata with mass conservation. Phys. A **343**, 185–200 (2004)

24. Suzuki, H., Ono, N., Yuta, K.: Several necessary conditions for the evolution of complex forms of life in an artificial environment. Artif. Life **9**(2), 153–174 (2003)

25. Tibshirani, R., Walther, G., Hastie, T.: Estimating the number of clusters in a data set via the gap statistic. J. Roy. Stat. Soc.: Ser. B (Stat. Methodol.) **63**(2), 411–423 (2001)

26. Varela, F.G., Maturana, H.R., Uribe, R.: Autopoiesis: the organization of living systems, its characterization and a model. Biosystems **5**(4), 187–196 (1974)

Emulation Games

See and Be Seen, an Subjective Approach to Analog Computational Neuroscience

Augusto Zubiaga$^{(\boxtimes)}$ (iD) and Lourdes Cilleruelo (iD)

University of the Basque Country UPV/EHU,
Campus Sarriena s/n, 48940 Leioa, Spain
{augustopedro.zubiaga,lourdes.cilleruelo}@ehu.eus

Abstract. Emulation consists in imitating a thing, trying to equal or even improve it. Playing means doing something for fun and entertainment. Under these premises, this paper proposes to describe an emulation game that addresses an approach to the area of *Computational Neuroscience*. It is in this context where we are designing, performing and valuing certain technical devices that could reveal, or make effective, computational events associated with the emergence of consciousness. The conceptual framework, in which is situated our game, can be defined as a possible epistemogony, a space where the physiology of knowledge and the making are activated, and from which we intend to highlight certain conditions that make possible both the so-called scientific research, and art making.

Keywords: Emulation game · Computational Neuroscience · Epistemogony

1 Introduction

Emulation consists in imitating a thing, trying to equal or even improve it. Playing means doing something for fun and entertainment. Under these premises, this paper proposes to describe an emulation game that addresses an approach to the area of Computational Neuroscience. It is in this context where we are designing, performing and valuing certain technical devices that could reveal, or make effective, computational events associated with the emergence of consciousness. The conceptual framework, in which is situated our game, can be defined as a possible epistemogony[1], a

---

[1] The epistemogony corresponds to the dimension in charge of the instances that make possible, support or justify the epistème (as sure and certain knowledge that the sciences try to reach) [1]. For example: "Just as a cosmogony may generate a class of theories of cosmology, or Hesiod's Theogony generated the class of theologies that are collectively called classical Greek mythology, so an epistemogony may be said to generate a class of epistemologies (e.g., Popper, Kuhn, Lakatos). For example, scientific realism is an epistemogony. The comparative examination of Popper, Kuhn, and Lakatos reveals that an epistemogony has three components: a logic of hypothesis generation, a logic of hypothesis testing (i.e., a logic of the process of confirmation), and a logic of generating consequences from (dis)confirmation. In any epistemogony, each of these may be either inductive or deductive. Under scientific realism, at least two of the three must be deductive." [2].

© Springer Nature Switzerland AG 2020
J. Romero et al. (Eds.): EvoMUSART 2020, LNCS 12103, pp. 210–225, 2020.
https://doi.org/10.1007/978-3-030-43859-3_15

space where the physiology of knowledge and the making are activated, and from which we intend to highlight certain conditions that make possible both the so-called scientific research, and art making.

The initial approach to this issue undertook a challenge to design an analogical electronic neuron[2], that is, to achieve a functional circuit with the materials and knowledge we have within our reach. After several unsatisfactory versions, and a continuous tinkering process, we found a possible configuration, following in the tradition of the earliest cybernetic systems[3].

Drawing inspiration from the twilight switches, it seemed a good idea to hybridize these circuits with optoelectronic devices. This possibility allowed to draw an analogy between the neuronal axon and an LED light source and to translate the excitatory and inhibitory synapses into LDR receivers. This is why the axons and receptors can be intercommunicated by optic fiber, simulating neural networks with complex dendritic ramifications. The capacity for automatic self-inhibition regulated by a discharge transistor, basically, is the key to making neuron operate smoothly. In simple terms, its activation provokes its inhibition in analogy to the well-known let me alone boxes, which, in our opinion, hide the deepest truth about the precise mechanism of the living beings (Figs. 1 and 2).

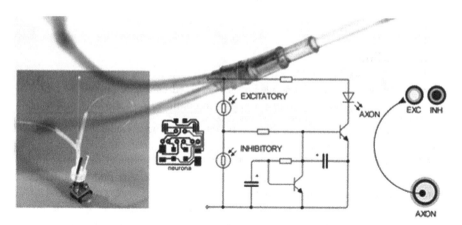

Fig. 1. Detail of dendritic branching made of optical fiber, an electronic neuron, photolithography for PCB development with SMD components; general circuit diagram, and normalized representation.

---

[2] Simplifying to the maximum we might say that the mechanism of a neuron consists of generating out pulses in response to input pulses, that is to say, exciting and reaching orgasm, or inhibiting and not be able to do so.

[3] Based on charge timers (resistance-capacitor).

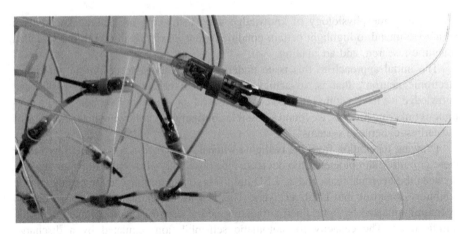

Fig. 2. Group of our electronic neuron emulators.

With the sufficient number of neurons, along with the ability to manufacture them economically, we began to investigate the effects of their connectivity. Thus, neural activity monitored and emulated in the Processing environment [3] has led to some empirical certainties.

2   Recognition

In Socratic terms, the task of knowledge is conceived as a recognition[4] of our ignorance. On this basis, the first empirical approach confirms the following properties:

- *Auto Excitation* - According to this configuration (see Fig. 3), a neuron connected to itself through its excitatory synapse, will generate a constant rhythmic pulse: it becomes a pacemaker. Although we are aware that a biological neuron is immeasurably more complex than this approach, we think that this first implementation provides sufficient functional plausibility to our device. However, the pulse of the living, in some way, beats in this configuration [5]. In addition, we note that an indeterminate number of independent neurons functioning as pacemakers, which are close enough to each other, tend to synchronize their beats, because of their structures are permeable to the influence of light variations of the immediate environment.
- *Self-Inhibition* - According to this configuration (see Fig. 3), a neuron connected to itself through its inhibitory synapse, will become refractory in relation to signals from other neurons.

---

[4] "Since for experimental science it is necessary to establish a radical distinction between the subject and the object of knowledge, the knowledge that it is able to provide is not and cannot be recognition as Socrates suggested, because what experimental science refuses to contemplate is the subject willing to know" [4, p. 166]. See also Erwin Schrödinger [10].

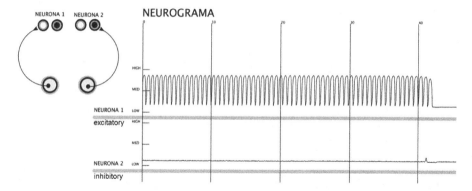

Fig. 3. Auto excitation and self-inhibition neurogram

- *Mutual Excitation* - According to this configuration (see Fig. 4), two neurons interconnected through their excitatory synapses will beat in unison, and the sum of their pulses will generate a wave of equal frequency but double amplitude.

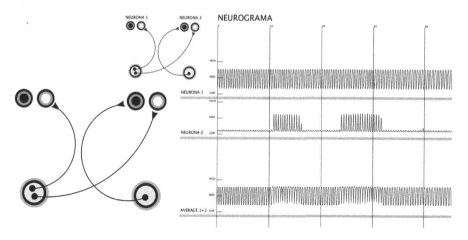

Fig. 4. Mutual excitation neurogram

- *Mutual inhibition* - Two neurons interconnected through their inhibitory synapses generate a synchronized oscillation pattern (see Fig. 5), in which the phase of each unit is displaced approximately 180° with respect to the other. The summation effect of the two pulses would generate a continuous and constant value pulse (HIGH).

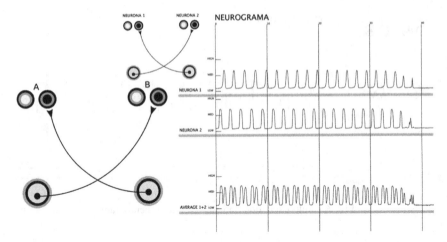

Fig. 5. Mutual inhibition neurogram

- *Excitation/inhibition* - If one neuron connects with another through an excitatory synapse (see Fig. 6), and the excitation of the second one is led to the inhibitory synapse of the first one, there is a mutual negation effect, in which the activation cycles of the two neurons are mutually neutralized by being in counter-phase. The result is a pulse output (LOW) close to zero, similar to a fibrillation effect.

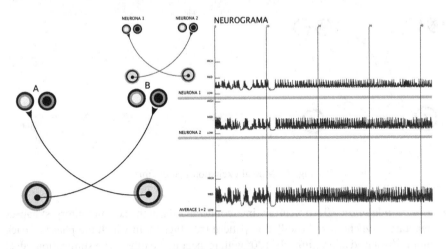

Fig. 6. Excitation/inhibition neurogram

In the following example (see Fig. 7), an oscillating arrangement of two neurons (A and C), which incorporates an interrupting effect from a third neuron (B) connected to a tactile-type sensing device, can be observed. As long as there is no tactile stimulus, the oscillator can work stably and reliably. In addition, if the sensitive neuron is excited, the oscillation effect is consequentially disturbed:

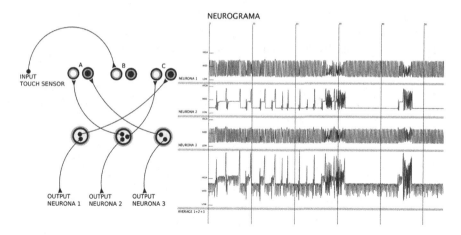

Fig. 7. Oscillator neurogram

These basic agonistic/antagonistic effects described can give us an idea of the immense complexity inherent in the ideation of a neuronal connectivity grammar that aspires to emulate complex behaviors without the aid of symbolic computation. However, that seems to be how life works... thanks to the help of huge amounts of time, and perhaps something else [6]. For example, the last arrangement would be useful for managing a biped locomotor system [12], which could unbalance its bilateral symmetry in order to modify its direction, because of an input in Neuron 2.

The fact that some fundamental operating principles for our neuron emulators were established opened up different possible avenues for research. The first approach can be defined as deductive, and the second, inductive.

3 Perspectives

3.1 Approach 1 (Deductive)

In an attempt to reconcile biologically inspired reactivities with binary symbolic computation, we can try to reproduce with our neurons the connective characteristics of the so-called logic gates (and, or, not, etc.), and emulate a syntax based on the implementation of procedures inspired by Boolean algebra. As shown in the following figures, in this sense, we have begun to approach to possible neural configurations that could reproduce the binary logic gates, with interesting results, but perhaps not completely comparable or equivalent, as one would expect.

Although, in these terms, we doubt that a formally solid complex grammar could be developed we nevertheless believe that the results achieved are significant on a heuristic framework. In this sense, these results have driven the focus of the research to look for patterns in an essentially non-deterministic environment. This has provided interesting -and repeatable- configurations of neuronal connectivity based on figures with a high degree of symmetry, which could perhaps be used to, for example, reverse-engineer neuromotor reactivities in the biological sphere [12].

The provisional results of our reinterpretation of the logic gate are shown below.

Logic Function AND

The graph (see Fig. 8) shows the effect of two touch sensors connected to the exciter synapses of Neuron 1 and Neuron 2. The stimulation of only one of these is not enough to excite the response of Neuron 3.

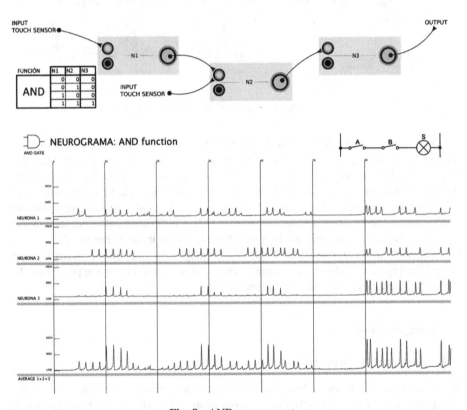

Fig. 8. AND arrangement

The AND function can be implemented by modulating the trigger weights, so that only the sum of Neurons 1 and 2 discharges reaches the excitation threshold of Neuron 3. As indicated in the graph, the weights of the synapses have been adjusted by means of moving the fiber optic terminals closer or farther away from the LDRs and LEDs, until we find the appropriate proportion. Thus, we reproduce the functioning of the so-called Perceptron[5].

---

[5] Wikipedia, https://es.wikipedia.org/wiki/Perceptr%C3%B3n, last accessed 2019/11/11.

Logic Function OR

The graph shows (see Fig. 9) two touch sensors connected to Neurons 1 and 2. Either of them indistinctly excite Neuron 3. The sum of the combined activity of the 3 neurons can be seen in the graph below.

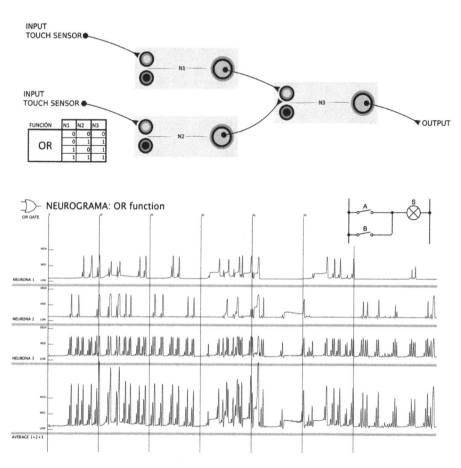

Fig. 9. OR arrangement

Logic Function NOT

The graph shows (see Fig. 10) a tactile sensor that excites Neuron 1 in turn connected to the inhibitory synapse of Neuron 2. The resulting effect consists of a reflexive activation of Neuron 2, at the moment when the signal from Neuron 1 ceases. The temporal sequence of activations of Neurons 1 and 2 can be seen in the graph below. This mechanism is identical to that of double inhibition.

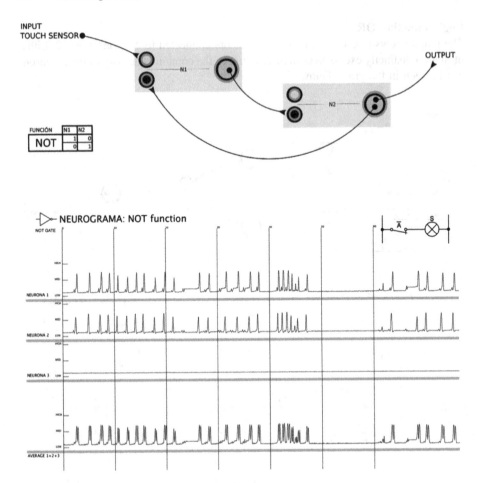

Fig. 10. NOT arrangement

Logic Function NOR

The graph shows (see Fig. 11) the activation sequence of 4 Neurons according to the NOR connective configuration. In the average, an ordered sequential activation can be observed which is only occasionally disturbed by the signal coming from a touch sensor connected to the exciter synapse of Neuron 1. For example, this configuration could be related to a synchronized activation sequence of four motor extremities.

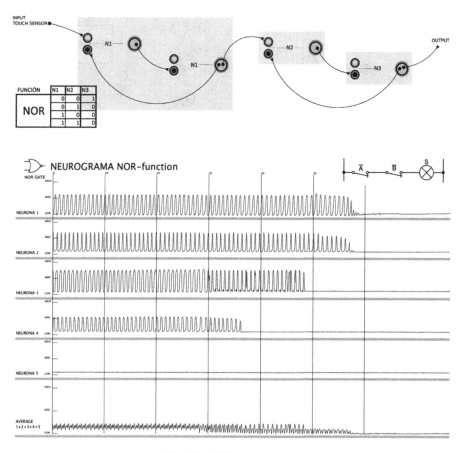

Fig. 11. NOR arrangement

Logic Function NAND

The graph shows (see Fig. 12) 4 Neurons interconnected according to a configuration associated with the NAND function. In the case reproduced in the neurogram in Fig. 12, N1 and N2 oscillate synchronously according to the Not configuration (Neurons 1, 2, 3 and 4). The result is an almost complete inhibition of N3 (Neurone 5), similar to a fibrillation effect.

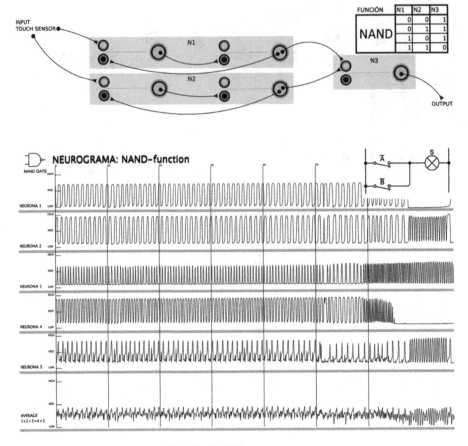

Fig. 12. NAND arrangement

Logic Function EXOR

The graph (see Fig. 13) shows 5 Neurons and 2 tactile sensors. Neuron 5 is an output dependent on simultaneous or alternating activation of the tactile inputs that excite neurons 1 and 3. Excitation of Neuron 3 inhibits the triggering of Neuron 5, until activation of the tactile sensor of Neuron 1 reverses the inhibitory effect of Neuron 3. In the absence of stimuli from Neurons 1 and 3, neurons 2 and 4 oscillate synchronously, generating a double frequency rhythm in Neuron 5.

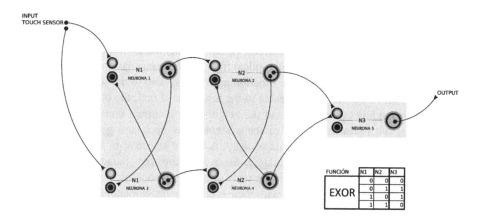

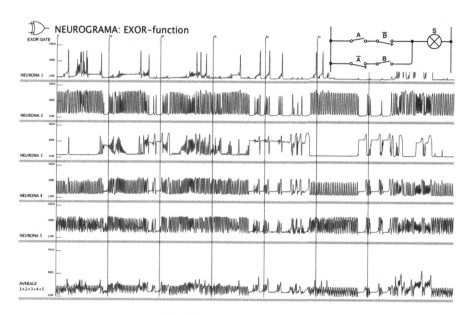

Fig. 13. EXOR arrangement

Discussion. A strictly structural reason for the difficult adaptation of these devices to the assumptions of formal logic would be that our neuron emulators, as biological neurons, do not exactly work by means of constant energy flows that can be channeled by opening or closing logic gates [11], but pulses and discrete loading and unloading times. However, since they are subject to great indeterminacy, they generate fields with a high degree of uncertainty (chaotic or virtually infinite states) [7].

This is due to the analogical architecture itself, based on timed cycles of charge and discharge of capacitors through variable resistances dependent on light, which transduce luminous fluxes in electrical current. A basic structural condition must be also added to this, which is the use of bare optical fiber for the transmission of information. This last feature entails exposing the functioning of the system to the surrounding light environment that is making it permeable to the influence of both the photonic particles from the surrounding environment and the neighboring neuronal units.

Therefore, the intrinsic activity of the system feeds back and generates internal filtrations and resonances and, simultaneously, is highly sensitive to variations in the environment. This apparent disadvantage is for us an interesting characteristic, because we believe that it brings it closer to the real functioning mode of biological nervous systems, which make it intertwined with their ecosystem, as emulators and resonators. In this context, certain feedback phenomena coming from the electromagnetic fields generated by the neuronal activity itself, which perhaps have something to do with the emergence of what is called the mind [8], should not be discarded.

Perhaps, a valid approach to a predictive understanding of the functioning of our system, which does not seem properly to be deterministic, should be of a statistical type1 [9]. In order to provide some predictability to the logical figures deployed a possible alternative could be to arrange them in massively parallel (redundant) structures.

3.2   Approach 2 (Inductive)

On the other hand, we could construct emulators of simple biological entities, constructing an artificial thought that would not be directly conditioned by presuppositions of symbolic logic -which would surely converge in something like a specifically human artificial intelligence-, but in the reproduction of what could be called patterns of conduct, not necessarily human.

In philosophical terms, on the one hand, it could be said that the first mode of approach, from top to bottom, is inspired by a Cartesian approach, since it attributes to the language structure the capacity to produce thought. A thought that will be logical, or will not be so, in the sense that it will be obliged to reproduce or deduce itself from its own presuppositions, in a certain tautological way [10]. On the other hand, the second mode (from bottom to top, nihilistic), starts from the multiple and indeterminate, entering into an uncertainty that promises nothing and takes nothing for granted. It suspected that the second mode is the one according to which the evolutionary principles have acted, which have given rise, among others, to specifically human thought, but also to other minds [8].

It is possible that there are no logical gates in logical thought, although there are in artificial intelligence... In this regard, we echo the reflection [9] according to which the human brain is capable of functioning as a computer. In fact, the human being is capable of using the tools of formal logic and mechanical thinking, language aided; but he is also capable of, as well as the rest of animals, thinking in other terms.

Under these premises, different theoretical-practical evolutionary developments in the form of works in process have been developed: on the one hand, aesthetic reflections around technologically mediated creative processes, and on the other, emulation games around the possibilities offered by analogical electronics based on the emulation of the functioning of biological entities (BEAM robotics). These researches have crystallized in art projects such as *See and be Seen* (Fig. 14 and 15) an artificial nervous system that has a vision device with which it accesses its own world.

See and Be Seen. The art installation *See and Be Seen* can be understood as an approach to the issue of the emergence of consciousness. The art installation consists in a mechanistic model that attempts to emulate the functioning of the nervous systems of biological entities, avoiding the use of computer models based on symbolic algorithms. On the contrary, it addresses the issue of emergence of consciousness from similarity, starting from functional units (analogue electronic neurons) that are interconnected in order to create networks that are capable of generating their own reality, and feeding on it.

The visual system has been emulated by arranging a surface with fiber optic terminals that functions as a retina, where a lens that emulates the crystalline lens focus the image. The image captured by the retina is sent through these fiber optic terminals, to the excitatory synapses of a group of neurons (32), which are radially arranged. Each of these 32 neurons makes inhibitory synapses with each contiguous neuron. This connective arrangement generates a constant signal, a basal pattern of activity (basal wave) that is disturbed by the variations in luminosity captured by each of the sensitive terminals of the retina.

In turn, the axons of the retinal terminals selectively excite 32 other neurons, organized in groups or ganglia of 3 and 4 interconnected neurons; which make up these groups form a kind of central nervous system, which is an internal correlate of the retinal structure, and that generates its own waves of activity. The interference between the flows coming from the retina and the internal states, through the optic nerve, generates new disturbances in the activity of the central ganglia. These disturbances projected to the outside, constitute a changing reality that in turn is perceived by the visual system.

The thinking machine feels, imagines and builds its own reality in a continuous feedback that can only be interrupted with our interposition as the real. The neural activity monitoring related to machine suggests the possibility of imagining some kind of Turing-Test that is trying to decide whether the vital signs, which are shown by sensors, correspond with the genuine activity of an authentic biological entity or not. When applicable, if the presence and nature of some kind of consciousness should be suspected, or on the contrary, it is just a simulation, and there is no ghost in the machine. Presence or representation? According to McFadden and Al-Khalili [6, p. 363], "the findings suggest that the brain's own electromagnetic field, generated by the firing of the nerves, influences on the firing itself, providing a kind of self-referential loop that many theorists claim to be an essential component of consciousness".

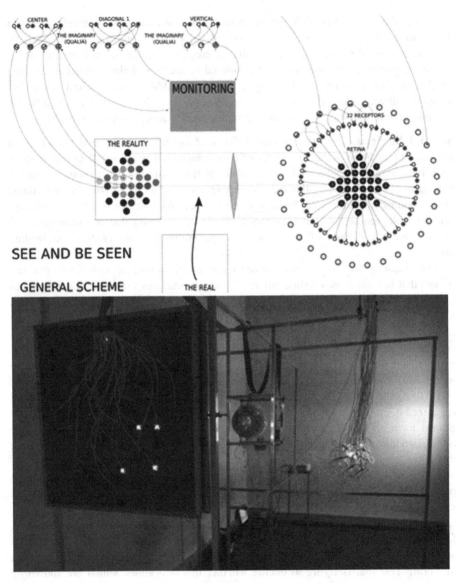

Fig. 14. A general scheme of *See and Be Seen* art installation

Fig. 15. A detail of the vision system of the *See and Be Seen* art installation

References

1. Moraza, J.L.: Arte y Saber. In: ARTELEKU. http://2003.arteleku.net/arteleku/program/archive/arte-y-saber. Accessed 11 Nov 2018
2. Cutler, R.M.: Complexity science and knowledge-creation in international relations theory. In: Institutional and Infrastructural Resources, Encyclopedia of Life Support Systems. EOLSS Publishers for UNESCO, Oxford (2002)
3. Processing Foundation: https://processingfoundation.org/. Accessed 11 Nov 2019
4. Ridao, J.M.: La democracia Intrascendente. Galaxia Gutenberg, Barcelona (2019)
5. Smith, C.U.M.: El Cerebro. Alianza Universidad, Madrid (1977)
6. McFadden, J., Al-Khalili, J.: Biología al límite. Cómo funciona la vida a muy pequeña escala. RBA, Barcelona (2019)
7. Baofu, P.: The Future of Complexity. Conceiving a Better Way to Understand Order and Chaos. World Scientific, Hackensack (2007)
8. Godfrey-Smith, P.: Otras Mentes. Taurus, Madrid (2017)
9. Nucleus Ambiguous: https://nucambiguous.wordpress.com/2014/11/22/the-neural-signal-and-the-neural-noise/. Accessed 10 Nov 2019
10. Schrödinger, E.: Mente y Materia. Tusquets, Barcelona (2018)
11. Smart, A.: Más allá de ceros y unos. Clave Intelectual, Madrid (2018)
12. Selverston, A.I.: Invertebrate central pattern generator circuits. The Royal Society, London (2010). https://doi.org/10.1098/rstb.2009.0270

References

Author Index

Printed in the United States
By Bookmasters